Spectroscopic Methods in Organic Chemistry

Sixth Edition

Spectroscopic Methods in Organic Chemistry

Sixth Edition

Dudley H. Williams, MA, PhD, ScD, FRS
Fellow of Churchill College, Cambridge

Ian Fleming, MA, PhD, ScD, FRS
Fellow of Pembroke College, Cambridge

McGraw-Hill Higher Education

London   Boston   Burr Ridge, IL   Dubuque, IA   Madison, WI   New York   San Francisco
St. Louis   Bangkok   Bogotá   Caracas   Kuala Lumpur   Lisbon   Madrid   Mexico City
Milan   Montreal   New Delhi   Santiago   Seoul   Singpaore   Sydney   Taipei   Toronto

Spectroscopic Methods in Organic Chemistry, Sixth Edition
Dudley H. Williams and Ian Fleming
ISBN-13 978-0-07-711812-9
ISBN-10 0-07-711812-X

**McGraw-Hill
Higher Education**

Published by McGraw-Hill Education
Shoppenhangers Road
Maidenhead
Berkshire
SL6 2QL
Telephone: 44 (0) 1628 502 500
Fax: 44 (0) 1628 770 224
Website: **www.mcgraw-hill.co.uk**

British Library Cataloguing in Publication Data
A catalogue record for this book is available from the British Library

Library of Congress Cataloguing in Publication Data
The Library of Congress data for this book has been applied for from the Library of Congress

New Editions Editor: Catriona Watson
Editorial Assistant: Katy Hamilton
Marketing Manager: Vanessa Boddington
Head of Production: Beverley Shields

Text Design by Ian Fleming
Cover design by Ego Creative
Printed and bound in the UK by Bell and Bain Ltd, Glasgow

First Edition published in 1966 by McGraw-Hill Publishing Company Limited
Second Edition published in 1973 by McGraw-Hill Book Company (UK) Limited
Third Edition published in 1980 by McGraw-Hill Book Company (UK) Limited
Fourth Edition published in 1987 by McGraw-Hill Book Company (UK) Limited
Fifth Edition published in 1995 by McGraw-Hill Education (Europe)

ISBN-13 978-0-07-711812-9
ISBN-10 0-07-711812-X

Contents

Chapter 3: Nuclear magnetic resonance spectra

Chapter 4: Mass spectra

Preface

This book is the sixth edition of a well-established introductory guide to the interpretation of the ultraviolet, infrared, nuclear magnetic resonance and mass spectra of organic compounds. It is a textbook suitable for a first course in the application of these techniques to structure determination, and as a handbook for organic chemists to keep on their desks throughout their career.

These four spectroscopic methods have been used routinely for several decades to determine the structure of organic compounds, both those made by synthesis and those isolated from natural sources. Every organic chemist needs to be skilled in how to apply them, and to know which method works for which problem. In outline, the ultraviolet spectrum identifies conjugated systems, the infrared spectrum identifies functional groups, the nuclear magnetic resonance spectra identify how the atoms are connected, and the mass spectrum gives the molecular formula. One or more of these techniques nowadays is very frequently enough to identify the complete chemical structure of an unknown compound, or to confirm the structure of a known compound. If they are not enough on their own, there are other methods that the organic chemist can turn to: X-ray diffraction, microwave absorption, the Raman spectrum, electron spin resonance and circular dichroism, among others. Powerful though they are, these techniques are all more specialised, and less part of the everyday practice of most organic chemists.

We have kept discussion of the theoretical background to a minimum, since application of the spectroscopic methods is possible without a detailed command of the theory behind them. We have described instead how the techniques work, and how to read each of the four kinds of spectra, including each of the most important 2D NMR spectra. We have included many tables of data at the ends of Chapters 2, 3 and 4, all of which are needed in the day-to-day interpretation of spectra. Finally in Chapter 5, we work through 11 examples of the way in which the four spectroscopic methods can be brought together to solve fairly simple structural problems, and there are 33 problem sets for you to work through for practice.

In preparing a sixth edition, we have almost completely rewritten the book, to reflect our experience teaching the subject, and to respond to changes that have taken place, both of emphasis and of fact, since the fifth edition was published. The chapters on UV and IR spectra are more concise, the chapter on NMR is expanded, and the chapter on MS made more specific to the everyday, rather than to the more specialised, applications of this technique. The appearance of IR absorptions, formerly gathered at the end of the chapter, are now illustrated at the relevant points in the text. Conversely, we have moved the tables of IR data to the end of the chapter, where they are more convenient for reference, and match the arrangement we have always used for the NMR and MS chapters. Most significantly, we have replaced all of the 60 MHz spectra used hitherto to explain the fundamentals of NMR spectroscopy with new and carefully chosen examples at 400 MHz or more. We have also chosen several new compounds with which to illustrate better the common 2D NMR techniques. We have added a number of tables of basic NMR information—chemical shift and coupling constants—for the most common nuclei other than ^1H and ^{13}C; namely ^{11}B, ^{15}N, ^{19}F, ^{29}Si and ^{31}P. Finally, all the spectra in the examples and problems in Chapter 5 have been redrawn, smaller, but with no loss of detail. In this way, all the spectra for each problem can be presented conveniently in portrait orientation, and on one page in most cases.

We have been helped by a number of colleagues whom we should like to thank: Dr Richard Horan for several of the samples used in the early pages of Chapter 3, Chris Jones for preparing the ester **1** in Chapter 3, a new compound designed to present the ideas of chemical shift without the complication of proton-proton coupling, and yet to have each of the four different levels of carbon substitution, Duncan Howe, who ran most of the 1D NMR spectra, Dr Nick Bampos who trained IF in the use of XWINNMR, Dr Ed Anderson who ran most of the 2D spectra, Ed Houghton and Elaine Stephens for advice about Mass Spectrometry as it is practised today, and Derek Pert at Bruker Spectrospin Ltd for the NMR spectra in Chapter 5. Ian Fleming would also like to acknowledge the assistance of XWINPLOT, Photoshop and ChemDraw in producing and editing all the figures in this edition.

Ian Fleming
Dudley H. Williams
Cambridge

Custom Publishing Solutions:
Let us help make
our *content* your *solution*

McGraw-Hill Education our aim is to help the lecturer find the most suitable content for their needs and most appropriate way to deliver the content their students. Our **custom publishing solutions** offer the best combination of content delivered in the way which suits lecturer and students the best.

The idea behind our custom publishing programme is that via a database of over two million pages called Primis, www.primisonline.com the lecturer can select just the material they wish to deliver to their students:

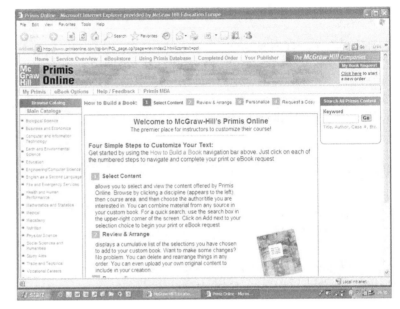

Lecturers can select chapters from:
- textbooks
- professional books
- case books and Taking Sides debate materials

Across the following imprints:
- McGraw-Hill Education
- Open University Press
- US and European material

There is also the option to include material authored by lecturers in the custom product – this does not necessarily have to be in English.

We will take care of everything from start to finish in the process of developing and delivering a custom product to ensure that lecturers and students receive exactly the material they need in the most appropriate format.

With a Custom Publishing Solution, students enjoy the best selection of material deemed to be the most suitable for learning everything they need for their courses – something of real value to support their learning. Teachers are able to use exactly the material they want, in the way they want, to support their teaching on the course.

Please contact **your local McGraw-Hill representative** with any questions or alternatively contact Warren Eels **e: warren_eels@mcgraw-hill.com**

Also Available from McGraw-Hill

General Chemistry

Burdge and Change, *Chemistry*
ISBN 13: 9780071102247, ISBN 10: 0071102248

Chang: *Chemistry, Ninth Edition*
ISBN13: 9780071105958, ISBN10: 0071105956

Laird, *University Chemistry*
ISBN13: 9780071287746, ISBN10: 0071287744

Silberberg, *Chemistry: The Molecular Nature of Matter and Change, Fifth Edition*
ISBN13: 9780071283540, ISBN10: 0071283544

Organic Chemistry

Smith, *Organic Chemistry, Second Edition*
ISBN13: 9780071286657, ISBN10: 0071286659

Carey, *Organic Chemistry, Seventh Edition*
ISBN13: 9780071102254, ISBN10: 0071102256

Physical Chemistry

Forthcoming in 2008: Levine, *Physical Chemistry, Sixth Edition*

1. Ultraviolet and visible spectra

1.1 Introduction

The ultraviolet and visible spectra of organic compounds are associated with transitions between electronic energy levels in which an electron from a low-energy orbital in the ground state is promoted into a higher-energy orbital. Normally, the transition occurs from a filled to a formerly empty orbital (Fig. 1.1) to create a singlet excited state. The wavelength of the absorption is a measure of the separation E of the energy levels of the orbitals concerned.

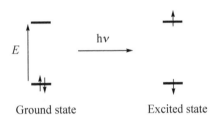

Fig. 1.1

Energy is related to wavelength by Eq. 1.1.

$$E(kJmol^{-1}) = \frac{1.19 \times 10^5}{\lambda(nm)} \tag{1.1}$$

Thus, 297 nm, for example, is equivalent to 400 kJ (~96 kcal)—enough energy to initiate many interesting reactions; compounds should not, therefore, be left in the ultraviolet beam any longer than is necessary.

1.2 Chromophores

The word chromophore is used to describe the system containing the electrons responsible for the absorption in question. Chromophores leading to the shortest wavelength absorption, in other words the highest energy separation, are found when electrons in σ-bonds are excited, giving rise to absorption in the 120-150 nm (1 nm = 10^{-7} cm = 10 Å = 1 mμ) range, corresponding to the transition x in Fig. 1.2. Isolated double bonds like that in ethene give rise to a strong absorption maximum at 162 nm, corresponding to the transition y in Fig. 1.2. Since the air is full of σ and π bonds, it strongly absorbs UV light below 200 nm, and this range is known as the vacuum ultraviolet, since air must be excluded from the instrument in order to detect the absorption. The absorption at these short wavelengths is difficult to measure and of little use in structure determination.

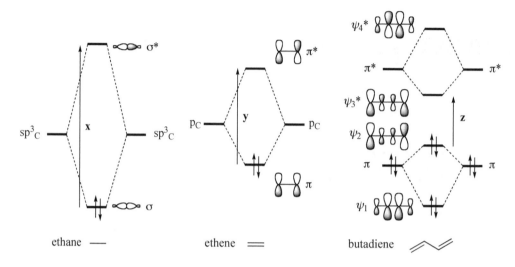

Fig. 1.2

Above 200 nm, however, excitation of electrons from conjugated π-orbitals, gives rise to readily measured and informative spectra. When two double bonds are conjugated, the energy level of the highest occupied molecular orbital ψ_2 (the HOMO) is raised in energy relative to the π orbital of the isolated double bonds, and that of the lowest unoccupied molecular orbital ψ_3^* (the LUMO) is lowered relative to π^*. The transition from ψ_2 to ψ_3^* is now associated with the even smaller value z in Fig. 1.2. This transition appears in the spectrum of butadiene as a strong, easily detected, and easily measured maximum at 217 nm. The same principle governs the energy levels when unlike chromophores, for example those of an α,β-unsaturated ketone, are conjugated together. Thus, methyl vinyl ketone has an absorption maximum at 225 nm, while neither a carbonyl group nor an isolated C=C double bond has a strong maximum above 200 nm. Since the longest wavelength absorption is usually that caused by promotion of an electron from the HOMO to the LUMO, it measures how far apart in energy those important orbitals are.

If yet another π-bond is brought into conjugation, the separation of the HOMO and LUMO is further reduced, and absorption occurs at a longer wavelength, with hexatriene absorbing at 267 nm. Each successive addition of a double bond reduces the energy gap, and moves the longest wavelength maximum further towards the visible. The long conjugated polyene lycopene, with 11 conjugated double bonds, has its longest wavelength absorption maximum at 504 nm (ε 158 000) with a tail reaching far into the visible region (absorbing the light in the blue to the orange range). Lycopene is responsible for the red colour of tomatoes. The most important point to be made is that, in general:

> **The longer the conjugated system, the longer the**
> **wavelength of the absorption maximum.**

1.3 The absorption laws

Two empirical laws have been formulated about the absorption intensity. Lambert's law states that the fraction of the incident light absorbed is independent of the intensity of the source. Beer's law states that the absorption is proportional to the number of absorbing molecules. From these laws, the remaining variables give the Eq 1.2.

$$\log_{10} \frac{I_0}{I} = \varepsilon.l.c \qquad (1.2)$$

I_0 and I are the intensities of the incident and transmitted light respectively, l is the path length of the absorbing solution in centimetres, and c is the concentration in moles per litre. $\log_{10}(I_0/I)$ is called the absorbance or optical density; ε is known as the molar extinction coefficient and has units of 1000 cm^2 mol^{-1} but the units are, by convention, not normally expressed.

1.4 Measurement of the spectrum

The ultraviolet or visible spectrum is usually taken using a dilute solution. An appropriate quantity of the compound (often about 1 mg when the compound has a molecular weight of 100-400) is weighed accurately, dissolved in the solvent of choice (see below), and made up to, for instance, 100 ml. A portion of this solution is transferred to a silica cell 1 cm from front to back internally (the value l in Eq. 1.2), and the pure solvent is transferred into an accurately matched cell. Two equal beams of ultraviolet or visible light are passed, one through the solution of the sample, and one through the pure solvent. The intensities of the transmitted beams are then compared over the whole wavelength range of the instrument. The spectrum is plotted automatically on most instruments as a $\log_{10}(I_0/I)$ ordinate and λ abscissa and might look something like Fig. 1.3, which is the spectrum of styrene (molecular weight 104) as a solution of 0.535 mg in 100 ml of hexane and a path length of 1 cm. For publication and comparisons the optical density is

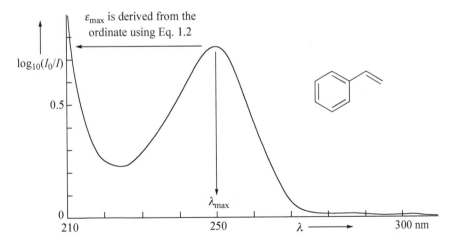

Fig. 1.3

converted to an ε versus λ or log ε versus λ plot using Eq. 1.2. The unit of λ is almost always nm. The intensity of a transition is better measured by the area under an absorption peak (when plotted as ε against frequency), but for convenience, and because of the difficulty of dealing with overlapping bands, ε_{max}, the maximum intensity of the absorption, is adopted in everyday use. Spectra are quoted, therefore, in terms of λ_{max}, the wavelength at the maximum of the absorption peak read directly off the plot like that in Fig. 1.3, where it is 250 nm, and ε_{max} is 14 700 calculated from the value of $\log_{10}(I_0/I)$, which is 0.756 on the plot in Fig. 1.3.

1.5 Vibrational fine structure

The excitation of electrons is accompanied by changes in the vibrational and rotational quantum numbers so that what would otherwise be an absorption line becomes a broad peak containing all the vibrational and rotational fine structure. Because of interactions of solute with solvent molecules this is usually not resolved, and a smooth curve is observed like that illustrated in Fig. 1.3. In the vapour phase, in non-polar solvents, and with certain peaks (e.g. benzene with the 260 nm band), vibrational fine structure is sometimes resolved.

1.6 Choice of solvent

The solvent most commonly used is 95% ethanol (commercial absolute ethanol contains residual benzene, which absorbs in the ultraviolet). It is cheap, a good solvent, and transparent down to about 210 nm. Fine structure, if desired, may be revealed by using cyclohexane or other hydrocarbon solvents which, being less polar, have less interaction with the absorbing molecules. Table 1.1 gives a list of common solvents and the minimum wavelength from which they may be used in 1 cm cells.

Table 1.1 Some solvents used in ultraviolet spectroscopy

Solvent	Minimum wavelength for 1 cm cell, nm
Acetonitrile	190
Water	191
Cyclohexane	195
Hexane	201
Methanol	203
Ethanol	204
Ether	215
Dichloromethane	220
Chloroform	237
Carbon tetrachloride	257

The effect of solvent polarity on the position of maxima is discussed in Sec. 1.8.

1.7 Selection rules and intensity

The irradiation of organic compounds does not always give rise to excitation of electrons from any filled orbital to any unfilled orbital, because there are rules based on symmetry governing which transitions are allowed. The intensity of the absorption is therefore a function of the 'allowedness', or otherwise, of the electronic transition and of the target area able to capture the light. Equation 1.3 gives the relationship between these variables.

$$\varepsilon = 0.87 \times 10^{20} P.a \qquad (1.3)$$

where P is called the transition probability (with values from 0 to 1) and a is the target area of the chromophore in Å^2. With common chromophores having an area typically of the order of 10 Å^2, a transition of unit probability will have an ε value of 10^5, and longer chromophores will have values in excess of this. In practice, a chromophore with two double bonds conjugated together giving rise to absorption by a fully allowed transition will have ε values of about 10 000, while forbidden transitions (which in practice occur with low transition probabilities), will have ε values below 1000. The important point is that, in general:

The longer the conjugated system, the more intense the absorption.

There are many factors that affect the transition probability, but most important are the rules governing which transitions are allowed and which forbidden. These are a function of the symmetry and multiplicity both of the ground state and excited state orbitals concerned. A full theoretical picture is given in the books by Jaffe and Orchin and by Murrell, listed in the bibliography, but a simple knowledge of which of the commonly encountered transitions are allowed and which are forbidden is adequate for anyone using UV spectra simply to determine organic structures or to follow reaction kinetics. Thus, the important promotion of an electron from the HOMO of a linear conjugated system to the LUMO of the same system is allowed, and always leads to intense absorption. In contrast, two important forbidden transitions are the n→π* band near 300 nm of ketones, with ε values of the order of 10 to 100, and the benzene 260 nm band and its equivalent in more complicated systems, with ε values from 100 upwards. 'Forbidden' transitions like these, with ε_{max} typically less than 1000, are observed because the symmetry which makes absorption strictly forbidden is broken by molecular vibrations or by the presence of unsymmetrical substitution. Both types are discussed further under the sections on ketones and aromatic systems.

1.8 Solvent effects

π→π*. The Frank-Condon principle states that during an electronic transition atoms do not move. Electrons, however, including those of solvent molecules, may reorganise. Most transitions result in an excited state more polar than the ground state; the dipole-dipole interactions with solvent molecules will, therefore, lower the energy of the excited state more than that of the ground state. Thus, it is usually observed that ethanol solutions give longer wavelength maxima than do hexane solutions. In other words, there is a small red shift of the order of 10-20 nm in going from hexane as solvent to ethanol.

n→π*. The weak transition of the oxygen lone pair in ketones—the n→π* transition—shows a solvent effect in the opposite direction. The solvent effect is now a result of the lesser extent to which solvents can hydrogen bond to the carbonyl group in the excited state than in the ground state. In hexane solution, for example, the absorption maximum of acetone is at 278 nm ($\varepsilon = 15$), whereas in aqueous solution the maximum is at 264.5 nm. The shift in this direction is known as a blue shift.

1.9 Searching for a chromophore

There is no easy rule or set procedure for identifying a chromophore—too many factors affect the spectrum and the range of structures that can be found is too great. The examination of a spectrum with particular regard for the following points is the first step to be taken.

The complexity and the extent to which the spectrum encroaches on the visible region. A spectrum with many strong bands stretching into the visible shows the presence of a long conjugated or a polycyclic aromatic chromophore. A compound giving a spectrum with one band (or only a few bands) below about 300 nm probably contains only two or three conjugated units.

The intensity of the bands, particularly the principal maximum and the longest wavelength maximum. This observation can be very informative. Simple conjugated chromophores such as dienes and α,β-unsaturated ketones have ε values of 10 000-20 000. The longer simple conjugated systems have principal maxima (usually also the longest wavelength maxima) with correspondingly higher ε values. Low intensity absorption bands in the 270-350 nm region, on the other hand, with ε values of 10-100, are the result of the n→π* transition of ketones. In between these extremes, the existence of absorption bands with ε values of 1000-10 000 almost always shows the presence of an aromatic system. Many unsubstituted aromatic systems show bands with intensities of this order of magnitude, the absorption being the result of a transition with a low transition probability, low because the symmetry of the ground and excited states make the transition forbidden. When the aromatic nucleus is substituted with groups that can extend the chromophore and break the symmetry, strong bands with ε values above 10 000 appear, but bands with ε values below 10 000 are often still present.

Confidence in the purity of the sample. It is always possible that weak bands are caused by small amounts of intensely absorbing impurities. Before any confidence can be put on an absorption with a low ε value, one must be sure of the purity of the sample.

Having made these observations, one should search for a model system which contains the chromophore and therefore gives a similar spectrum to that which is being examined. This may be difficult in rare cases; but so many spectra are now known, and the changes caused by substitution so well documented, that the task can be a simple one. The first tool which an organic chemist requires is a general knowledge of the simple chromophores and the changes which structural variations make in the absorption pattern. The remaining task, that of searching through the literature, is greatly facilitated by the existence of indexes and compilations. The major collection of data is *Organic Electronic Spectral Data*, Wiley, New York, Vols. 1-31 (1960-96). This most valuable collection has been prepared by a complete search of the major journals from 1945 until 1989, but has been discontinued since then. The compounds are indexed by their empirical formulae, and λ_{max} and $\log_{10}\varepsilon$ values are quoted together with literature references.

The search for a chromophore is likely to be assisted by the other and more powerful physical methods described in this book. The UV spectrum will mainly help to decide on the likely degree to which the functional groups are conjugated, and is often the last of the physical methods to be turned to. The range of structures in which a search must be made can be narrowed, for example, to aromatic compounds on the strength of infrared or ^{1}H NMR aromatic C—H absorptions. Similarly the presence of an α,β-unsaturated ketone may be inferred from the C=O stretching vibration observed in the infrared spectrum, the presence of a low-field carbon resonance in the ^{13}C NMR spectrum and then confirmed by an appropriate ultraviolet spectrum. One area where the UV spectrum can be especially important at an early stage is in the assignment of a structure to a natural product. These compounds, isolated from a natural source, have no history to help in the structure determination, in contrast to the products of a reaction between two known chemicals. The positive identification of a likely chromophore in a natural product can help to identify to which class the natural product belongs.

1.10 Definitions

The following words and symbols are commonly used:

Red shift or *bathochromic effect*. A shift of an absorption maximum towards longer wavelength. It may be produced by a change of medium or by the presence of an auxochrome.

Auxochrome. A substituent on a chromophore which leads to a red shift. For example, the conjugation of the lone pair on the nitrogen atom of an enamine shifts the absorption maximum from the isolated double bond value of 190 nm to about 230 nm. The nitrogen substituent is the auxochrome. An auxochrome, then, extends a chromophore to give a new chromophore.

Blue shift or *hypsochromic effect*. A shift towards shorter wavelength. This may be caused by a change of medium and also by such phenomena as the removal of conjugation. For example, the conjugation of the lone pair of electrons on the nitrogen atom of aniline with the π-bond system of the benzene ring is removed on protonation. Aniline absorbs at 230 nm (ε 8600), but in acid solution the main peak is almost identical with that of benzene, being now at 203 nm (ε 7500). A blue shift has occurred on protonation.

Hypochromic effect. An effect leading to decreased absorption intensity.

Hyperchromic effect. An effect leading to increased absorption intensity.

λ_{max}. The wavelength of an absorption maximum.

ε. The extinction coefficient defined by Eq. 1.2.

$E_{1cm}^{1\%}$. Absorption [$\log_{10} (I_0/I)$] of a 1% solution in a cell with a 1 cm path length. This is used in place of ε when the molecular weight of a compound is not known, or when a mixture is being examined, so that Eq. 1.2 cannot be used to define the intensity of the absorption.

Isosbestic point. A point common to all curves produced in the spectra of a compound taken at several pH values—the one point where the absorption intensity does not change as the pH changes.

1.11 Conjugated dienes

The energy levels of butadiene have been illustrated in Fig. 1.2. The transition z gives rise to strong absorption at 217 nm (ε 21 000). Alkyl substitution extends the chromophore, in the sense that there is a small interaction (called hyperconjugation) between the σ-bonded electrons of the alkyl groups and the π-bond system. The result is a small red shift with alkyl substitution, just as there is a red shift (though a relatively large one) in going from an isolated double bond to a conjugated diene or to an enamine.

The effect of alkyl substitution, in dienes at least, is approximately additive, and a few rules suffice to predict the position of absorption in open chain dienes and dienes in six-membered rings. Open chain dienes exist normally in the energetically preferred s-*trans* conformation, while homoannular dienes must be in the s-*cis* conformation. These conformations are illustrated in the part structures **1** (heteroannular diene) and **2** (homo-annular diene). It is not clear why, but the s-*cis* arrangement, as in the diene **2**, leads to longer wavelength absorption than does the s-*trans* arrangement in the diene **1**. Also, because of the shorter distance between the ends of the chromophore, s-*cis* dienes give maxima of lower intensity ($\varepsilon \sim$10 000) than the maxima of s-*trans* dienes ($\sigma \sim$20 000).

The rules for predicting the absorption of open chain and six-membered ring dienes were first made by Woodward in 1941, and were a breakthrough in showing that physical methods could be used in the details of structure determination. Since that time they have been modified by Fieser and by Scott as a result of experience with a larger number of dienes and trienes. The modified rules are given in Table 1.2.

Table 1.2 Rules for diene and triene absorption

Value assigned to parent s-*trans* diene (like **1**)	214 nm
Value assigned to parent s-*cis* diene (like **2**)	253 nm
Increment for:	
(a) each alkyl substituent or ring residue	5 nm
(b) the exocyclic nature of any double bond	5 nm
(c) a double-bond extension	30 nm
(d) auxochrome:	
—OAcyl	0 nm
—OAlkyl	6 nm
—SAlkyl	30 nm
—Cl, —Br	5 nm
—NAlkyl$_2$	60 nm
λ_{calc} Total	

(Reprinted with permission from A. I. Scott, *Interpretation of the Ultraviolet Spectra of Natural Products*, Pergamon Press, Oxford, 1964.)

For example, the diene **1** would be calculated to have a maximum at 234 nm by the following addition:

Parent value	214 nm
Three-ring residues (marked *a*) 3 × 5 =	15 nm
One exocyclic double bond (the Δ^4 bond is exocyclic to ring B)	5 nm
Total	234 nm

A typical value observed for a steroid with this part structure is 235 nm (ε 19 000).

By similar calculation, the diene **2** would be expected to have a maximum at 273 nm, and steroids like this typically have one at 275 nm. Though ethanol is the usual solvent, change of solvent has little effect.

There are a large number of exceptions to the rules, where special factors can operate. Distortion of the chromophore may lead to red or blue shifts, depending on the nature of the distortion.

Thus, the strained molecule verbenene **3** has a maximum at 245.5 nm, whereas the usual calculation gives a value of 229 nm. The diene **4** might be expected to have a maximum at 273 nm, but distortion of the chromophore, presumably out of planarity with consequent loss of conjugation, causes the maximum to be as low as 220 nm with a similar loss in intensity (ε 5500). The diene **5**, in which coplanarity of the diene is more likely, gives a maximum at 248 nm (ε 15 800), but it still does not agree with the expected value. Change of ring size in the case of simple homoannular dienes also leads to departures from the predicted value of 263 nm as follows: cyclopentadiene, 238.5 nm (ε 3400); cycloheptadiene, 248 nm (ε 7500); while cyclohexadiene is close at 256 nm (ε 8000). The lesson, an important one, is that when the ultraviolet spectrum of an unknown compound is to be compared with that of a model compound, then the choice of model must be a careful one. Allowance must be made for the likely shape of the molecule and for any unusual strain. Some general comments on the effect of steric hindrance to coplanarity are given in Sec. 1.24.

1.12 Polyenes

As the number of double bonds in conjugation increases, the wavelength of maximum absorption encroaches on the visible region. A number of subsidiary bands also appear and the intensity increases. Table 1.3 gives examples of the longest wavelength maxima of some simple conjugated polyenes showing these trends.

Table 1.3 Longest wavelength maxima of some simple polyenes

	trans-Me(CH=CH)$_n$Me		*trans*-Ph(CH=CH)$_n$Ph	
n	λ_{max}, nm	ε	λ_{max}, nm	ε
3	274.5	30 000	358	75 000
4	310	76 500	384	86 000
5	342	122 000	403	94 000
6	380	146 500	420	113 000
7	401	-	435	135 000
8	411	-	-	

The appearance of the spectra of some of these simple polyenes is illustrated in Fig. 1.4, which shows vividly the two main lessons enshrined in the boxes on pp. 2 and 5. In addition, in each spectrum, shorter wavelength absorption maxima are visible. They are the result of other transitions than just the HOMO-LUMO transition that is responsible for the longest wavelength maximum. The longer the conjugated system the more transitions become possible, and the pattern of the maxima is characteristic of the polyene concerned. It can be used as a kind of fingerprint, but in using UV spectra for structure determination we largely concentrate on the longest wavelength maximum, and the hint it gives us about the length of the conjugated system.

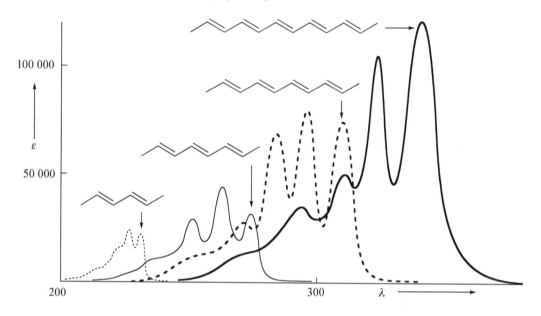

Fig. 1.4
(Replotted from P. Nayler and M. C. Whiting, *J. Chem. Soc.*, 1955, 3042.)

Several attempts, both empirical and theoretical, have been made to correlate quantitatively the principal or longest wavelength maximum with chain length. Some of the theoretical treatments have been based on the simple 'electron in the box' wave equation, in which the walls of the box are usually considered to be one average bond length beyond each end of the chromophore. Increasing values of λ_{max} are found for increasing length in a conjugated polyene, but quantitatively the correlation is less satisfactory. For example, the simple theory might indicate that, as the chain length increases, the value of λ_{max} for long chains would increase proportionately, whereas in practice there is a convergence, which can be seen already in Table 1.4. More sophisticated treatments, allowing for the variation in bond lengths between the double and single bonds, have been made and are described in Murrel's book. An interesting simplification is provided by the cyanine dye analogues **6** in which overlap leads to

$$Me_2N \quad \longleftrightarrow \quad Me_2N$$

6

uniform bond lengths and bond orders along the polyene chain. Calculations based on the 'electron in the box' lead in this case to values close to those observed: $\lambda_{max} = 309$ ($n = 1$), 409 ($n = 2$) and 511 ($n = 3$) nm.

In a long-chain polyene, change from a *trans* to a *cis* configuration at one or more double bonds lowers both the wavelength and the intensity of the absorption maximum as a result of steric problems in attaining coplanarity.

1.13 Polyeneynes and poly-ynes

The ultraviolet spectra of many natural polyeneynes and poly-ynes are known, and have been used in the elucidation of structure. They illustrate how a family of natural products can be detected and identified. A distinctive feature in the UV spectrum, when more than two triple bonds are conjugated, is a series of low-intensity 'forbidden' bands (ε ~100-200) at regular intervals of 2300 cm^{-1} (note the frequency units, frequency being directly proportional to energy whereas wavelength is not) together with high-intensity bands (ε ~10^5) at intervals of 2600 cm^{-1}. This characteristic spiky appearance of the spectra was helpful in screening crude plant extracts for acetylenic compounds. The principal maxima from each of these groups of lines are listed in Table 1.4, in which the trends closely resemble those for polyenes in Table 1.3.

In a representative application in structure determination, Fig. 1.5 shows the UV spectrum of dehydromatricaria ester **8**. In this compound, the longer wavelength bands are considerably more intense than in simple poly-ynes, but they are, like a fingerprint, characteristic of an enetriyne chromophore. The structure of a natural product **7** was assigned with the help of its UV spectrum, which showed a similar pattern, with each of

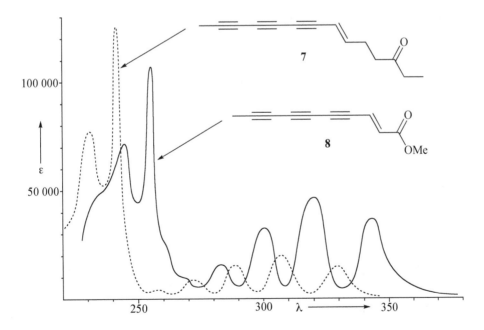

Fig. 1.5

(Replotted from J. S. Sörensen, T. Bruun, D. Holme and N. A. Sörensen, *Acta Chem. Scand.*, 1954, **8**, 28 and F. Bohlmann, H.-J. Mannhardt, and H. G. Viehe, *Chem. Ber.*, 1955, **88**, 361.)

Table 1.4 Principal maxima for conjugated poly-ynes $Me(C\equiv C)_nMe$

n	λ_{max}, nm	ε	λ_{max}, nm	ε
2	-	-	250	160
3	207	135 000	306	120
4	234	281 000	354	105
5	260.5	352 000	394	120
6	284	445 000	-	-

the peaks shifted to shorter wavelength. Thus, the UV spectrum indicated that the structure might be an enetriyne without the extra conjugation of the ester carbonyl group. When the particular enetriyne **7** was synthesised, it proved to be identical to the natural product. This is an example of the way in which an organic chemist deals with the comparison of UV spectra: the enetriyne chromophore of the ester **8** is present in the ketone **7**, and the latter therefore continues to show similar features to those of the former, with a blue shift because of the relatively shorter conjugated system.

Table 1.5 Rules for α,β-unsaturated ketone and aldehyde absorption in ethanol

ε values are usually above 10 000 and increase with the length of the conjugated system.

Value assigned to parent α,β-unsaturated six-ring or acyclic ketone		215 nm
Value assigned to parent α,β-unsaturated five-ring ketone		202 nm
Value assigned to parent α,β-unsaturated aldehyde		207 nm
Increments for (a) a double bond extending the conjugation		30 nm
(b) each alkyl group or ring residue α		10 nm
β		12 nm
γ and higher		18 nm
(c) lone-pair auxochromes (i) —OH	α	35 nm
	β	30 nm
	γ	50 nm
(ii) —OAc	α, β, γ	6 nm
(iii) —OMe	α	35 nm
	β	30 nm
	γ	17 nm
	δ	31 nm
(iv) —SAlk	β	85 nm
(v) —Cl	α	15 nm
	β	12 nm
(vi) —Br	α	25 nm
	β	30 nm
(vii) —NR$_2$	β	95 nm
(d) the exocyclic nature of any double bond		5 nm
(e) homodiene component		39 nm
$\lambda_{calc(EtOH)}$	Total	

(Reprinted with permission from A. I. Scott, *Interpretation of the Ultraviolet Spectra of Natural Products*, Pergamon Press, Oxford, 1964.)

1.14 Ketones and aldehydes; $\pi \rightarrow \pi^*$ transitions

As with dienes, Woodward formulated a set of rules for predicting the UV absorption of α,β-unsaturated ketones and aldehydes in ethanol. These rules, subsequently modified by Fieser and by Scott, are given in Table 1.5. For λ_{calc} in other solvents, a solvent correction from Table 1.6 must be subtracted from the above value, because the spectra are affected significantly by the solvent as a result of the change in polarity on excitation.

Table 1.6 Solvent corrections for α,β-unsaturated ketones

Solvent	Correction, nm
Water	−8
Ethanol	0
Methanol	0
Chloroform	+1
Dioxan	+5
Ether	+7
Hexane	+11
Cyclohexane	+11

(Reprinted with permission from A. I. Scott, *Interpretation of the Ultraviolet Spectra of Natural Products*, Pergamon Press, Oxford, 1964.)

For example, mesityl oxide ($Me_2C{=}CHCOMe$) may be calculated to have λ_{max} at $215 + (2 \times 12) = 239$ nm. The observed value is 237 nm (ε 12 600). A more complicated example, the trienone chromophore of **9**, would be calculated to have a maximum at 349 nm by the following addition.

9

Parent value	215 nm
β-Alkyl substituent (marked *a*)	12 nm
ω-Alkyl substituent (marked *b*)	18 nm
2 × Extended conjugation	60 nm
Homoannular diene component	39 nm
Exocyclic double bond (the α,β-double bond is exocyclic to ring A)	5 nm
Total	349 nm

The observed values of λ_{max} for a trienone with this substructure are 230 nm (ε 18 000), 278 nm (ε 3720) and 348 nm (ε 11 000). As was the case with simple polyenes, the long chromophore present in this example gives rise to several peaks, with the longest wavelength peak in good agreement with the prediction.

An important general principle is illustrated by the calculation for the cross-conjugated trienone **10**. In this case the main chromophore is the linear dienone portion, since the Δ^5-double bond is not in the longest conjugated system. The calculation, along the lines above, gives a value of 324 nm. The observed values are 256 nm and 327 nm. The former might be from the Δ^5-7-one system (λ_{calc} = 244 nm), but a positive identification of this sort in a complicated system is largely unjustified.

10   **11**

Certain special changes in structure, as noted in the case of dienes in Sec. 1.11, also lead to departures from the rules given above. The effect of the five-membered ring in cyclopentenones is accommodated in the rules; but when the carbonyl group is in a five-membered ring and the double bond is exocyclic to the five-membered ring, a parent value of about 215 nm holds. Another special case, verbenone **11**, would be calculated to have a maximum at 239 nm but actually has a maximum at 253 nm, an increment for strain of 14 nm, close to the increment for the corresponding diene **3**.

1.15 Ketones and aldehydes; n→π* transitions

Saturated ketones and aldehydes show a weak symmetry-forbidden band, in the 275-295 nm range (ε ~20), from excitation of an oxygen lone-pair electron (n = p_O) into the antibonding π* orbital of the carbonyl group, as shown on the left of Fig. 1.6.

Aldehydes and the more heavily substituted ketones absorb at the long wavelength end of this range. Electronegative substituents on the α-carbon atoms increase (when oriented like an axial substituent on C-2 in a cyclohexanone) or decrease (when equatorial) the wavelength. When the carbonyl group is directly attached to an electronegative element X—as in an ester, an acid, or an amide—the π* orbital is slightly raised in energy because the substituent, having a lone pair, is a π-donor. The n (p_O) level of the lone pair, on the other hand, is lowered because it is conjugated to the C—X bond, which is a σ-withdrawing group. The result is that the n→π* transition of these compounds is shifted to shorter wavelength into the relatively inaccessible 200-215 nm range. The presence, therefore, of a weak band in the 275-295 nm region is positive identification of a ketone or aldehyde carbonyl group (nitro groups show a similar band and, of course, impurities must be absent). In contrast, if the carbonyl group is directly attached to an electropositive substituent—as in an acylsilane where the silyl group is σ-donating and π-withdrawing—the π* orbital is lowered in energy, the n orbital is raised in energy and the n→π* transition is shifted to longer wavelength, close to 370 nm for saturated acylsilanes and 420 nm for aryl or α,β-unsaturated acylsilanes. These compounds are yellow and green, respectively, as a consequence of the tail of these absorptions reaching into the visible.

α,β-Unsaturated ketones show slightly stronger n→π* absorption (ε ~100) in the 300-350 nm range, since the ψ_3* orbital is lowered in energy by conjugation relative to the π* level of a simple carbonyl group, but the n level of the lone pair is largely unaltered, as shown on the right in Fig. 1.6. The precise position of these bands is not predictable from the extent of alkylation, but is a regular function of the conformation of γ-substituents,

substituents oriented so that they overlap with the π-system shifting the absorption to longer wavelength, since they extend the conjugation. The position and intensity of n→π* bands are also influenced by transannular interactions (see Sec. 1.23) and by solvent (see Sec. 1.8).

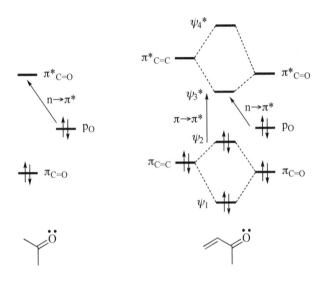

Fig. 1.6

The n→π* transitions of α-diketones in the diketo form give rise to two bands, one in the usual region near 290 nm (ε ~30) and a second (ε ~10-30) that stretches into the visible in the 340-440 nm region, and gives rise to the yellow colour of some of these compounds. (See also quinones in Sec. 1.21, quinones being α-, or vinylogous α-, diketones.)

1.16 α,β-Unsaturated acids, esters, nitriles and amides

α,β-Unsaturated acids and esters follow a trend similar to that of ketones but at slightly shorter wavelength. The rules for alkyl substitution, summarised by Nielsen, are given in Table 1.7. The change in going from acid to ester is usually not more than 2 nm.

Table 1.7 Rules for α,β-unsaturated acids' and esters' absorption (ε values are usually above 10 000)

β-Monosubstituted	208 nm
αβ- or ββ-disubstituted	217 nm
αββ-Trisubstituted	225 nm
Increment for:	
(a) a double bond extending the conjugation	30 nm
(b) the exocyclic nature of any double bond	5 nm
(c) when the double bond is endocyclic in	
a five-or seven-membered ring	5 nm
λ_{calc}	Total

α,β-Unsaturated nitriles absorb at wavelengths slightly lower than the corresponding acids, and α,β-unsaturated amides lower still, usually near 200 nm ($\varepsilon \sim 8000$). α,β-Unsaturated lactams have an additional band at 240-250 nm ($\varepsilon \sim 1000$).

1.17 The benzene ring

Benzene absorbs at 184 (ε 60 000), 203.5 (ε 7400) and 254 (ε 204) nm in hexane solution; it is illustrated by the dashed line in Fig. 1.7. The latter band, sometimes called the *B*-band, shows vibrational fine structure. Although a 'forbidden' band, it owes its appearance to the loss of symmetry caused by molecular vibrations; indeed, the 0→0 transition (the transition between the ground state vibrational energy level of the electronic ground state to the ground state vibrational energy level of the electronic excited state) is not observed.

When the aromatic ring is substituted by alkyl groups, for example, or is an aza analogue such as pyridine, the symmetry is lowered and the 0→0 transition is then observed, although the spectrum is little changed otherwise. The presence of fine structure resembling that shown in Fig. 1.7 is characteristic of the simpler aromatic molecules.

Table 1.8 Absorption maxima of substituted benzene rings Ph—R

R	λ_{max} nm (ε) (solvent H_2O or MeOH)					
—H	203.5	(7400)	254	(204)		
—NH_3^+	203	(7500)	254	(160)		
—Me	206.5	(7000)	261	(225)		
—I	207	(7000)	257	(700)		
—Cl	209.5	(7400)	263.5	(190)		
—Br	210	(7900)	261	(192)		
—OH	210.5	(6200)	270	(1450)		
—OMe	217	(6400)	269	(1480)		
—SO_2NH_2	217.5	(9700)	264.5	(740)		
—CN	224	(13 000)	271	(1000)		
—CO_2^-	224	(8700)	268	(560)		
—CO_2H	230	(11 600)	273	(970)		
—NH_2	230	(8600)	280	(1430)		
—O^-	235	(9400)	287	(2600)		
—NHAc	238	(10 500)				
—COMe	245.5	(9800)				
—$CH=CH_2$	248	(14 000)	282	(750)	291	(500)
—CHO	249.5	(11 400)				
—Ph	251.5	(18 300)				
—OPh	255	(11 000)	272	(2000)	278	(1800)
—NO_2	268.5	(7800)				
—$CH=CHCO_2H$	273	(21 000)				
—CH=CHPh	295.5	(29 000)				

(Most values taken with permission from H. H. Jaffe and M. Orchin, *Theory and Applications of Ultraviolet Spectroscopy*, Wiley, New York, 1962.)

When the benzene ring has a lone-pair or π-bonded substituent, in other words an auxochrome, the chromophore is extended. Quantitative prediction of the effects of such substituents is not as simple as it was with dienes and unsaturated ketones, but Sec. 1.18 gives an account of some of the trends observed with substituted benzene rings.

1.18 Substituted benzene rings

Table 1.8 gives the wavelength of absorption maxima in the spectra of a range of monosubstituted benzenes, showing how, as usual, the wavelength and intensity of the absorption peaks increase with an increase in the extent of the chromophore.

As more conjugation is added to the benzene ring, the band originally at 203.5 nm (sometimes called the K-band) effectively 'moves' to longer wavelength, and moves 'faster' than the B-band, which was originally at 254 nm, eventually overtaking it. This can be seen in the two other spectra recorded on Fig. 1.7: benzoic acid (the solid line) shows the K-band at 230 nm with the B-band still clearly visible at 273 nm; but with the longer chromophore of cinnamic acid (bold solid line) the K-band has moved to 273 nm and the B-band is completely submerged. In the latter case, we can see how the even stronger band, originally at 184 nm, has also moved, but has still not reached the accessible region. It is responsible for what is called end absorption, that is the long-wavelength side of an absorption peak, the maximum of which is below the range of the instrument.

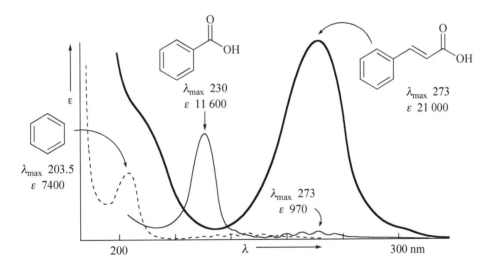

Fig. 1.7

In disubstituted benzenes, two situations are important. When electronically complementary groups, such as amino and nitro, are situated *para* to each other as in *p*-nitroaniline **12**, there is a pronounced red shift in the main absorption band, compared to the effect of either substituent separately, caused by the extension of the chromophore from the electron-donating group to the electron withdrawing group through the benzene ring, as symbolised by the curly arrows. Alternatively, when two groups are situated *ortho* or *meta* to each other, or when the *para* disposed groups are not complementary, as in *p*-dinitrobenzene **13**, then the observed spectrum is usually close to that of the separate, non-interacting, chromophores. These principles are illustrated by the examples in Table 1.9. The values in this table should be compared with each other and with the values for the single substituents separately given in Table 1.8.

12

λ_{max} 375 nm (ε 16 000)

13

λ_{max} 260 nm (ε 13 000)

In particular it should be noted that those compounds with non-complementary substituents, or with an *ortho* or *meta* substitution pattern, actually have a band (though a much weaker one) at longer wavelength than the compounds with interacting *para*-disubstituted substituents. This fact is not in accord with the simple resonance picture; neither is the similarity of the *ortho* to the *meta* disubstituted cases. This is another case in which molecular orbital theory (too complicated to be introduced here, but dealt with in Murrel's book) gives a better picture.

Table 1.9 Absorption maxima of disubstituted benzene rings

$$R^1—C_6H_4—R^2$$

R^1	R^2		λ_{max} (EtOH) nm (ε)				
—OH	—OH	*o*	214	(6000)	278	(2630)	
—OMe	—CHO	*o*	253	(11 000)	319	(4000)	
—NH$_2$	—NO$_2$	*o*	229	(16 000)	275	(5000)	405 (6000)
—OH	—OH	*m*	277	(2200)			
—OMe	—CHO	*m*	252	(8300)	314	(2800)	
—NH$_2$	—NO$_2$	*m*	235	(16 000)	373	(1500)	
—Ph	—Ph	*m*	251	(44 000)			
—OH	—OH	*p*	225	(5100)	293	(2700)	
—OMe	—CHO	*p*	277	(14 800)			
—NH$_2$	—NO$_2$	*p*	229	(5000)	375 (16 000)		
—Ph	—Ph	*p*	280	(25 000)			

In the case of disubstituted benzene rings in which the electron-donating group is complemented by an electron-withdrawing carbonyl group, some quantitative assessments may be made. These apply to the compounds RC_6H_4COX in which X is alkyl, H, OH, or OAlkyl, and refer to the strongest band in the accessible region, which is often the only measured band in the highly conjugated *para*-disubstituted systems. The calculation is based on a parent value with increments for each substituent. Polysubstituted benzene rings should be treated with caution, particularly when the substitution might lead to steric hindrance preventing coplanarity of the carbonyl group and the ring. Table 1.10 gives the rules for this calculation. In the absence of steric hindrance to coplanarity, the calculated values are usually within 5 nm of the observed values.

Table 1.10 Rules for the principal band of substituted benzene derivatives RC_6H_4COX

Parent chromophore: X	λ_{max} (EtOH) nm
alkyl or ring residue	246
H	250
OH or Oalkyl	230

Increment (nm) for each substituent: R	o, m or p	Increment	Increment (nm) for each substituent: R	o, m or p	Increment
Alkyl or ring residue	o, m	3	Br	o, m	2
	p	10		p	15
OH, OMe, Oalkyl	o, m	7	NH₂	o, m	13
	p	25		p	58
O⁻	o	11	NHAc	o, m	20
	m	20		p	45
	p	78	NHMe	p	73
Cl	o, m	0	NMe₂	o, m	20
	p	10		p	85

(Reprinted with permission from A. I. Scott, *Interpretation of the Ultraviolet Spectra of Natural Products*, Pergamon Press, Oxford, 1964.)

6-Methoxytetralone **14** provides an example:

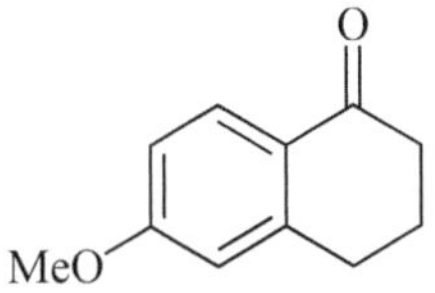

14

Parent value	246 nm
ortho alkyl	3 nm
para MeO	25 nm
λ_{calc}	274 nm

The maximum actually occurs at 276 nm (ε 16 500).

Other electron withdrawing groups, like cyano and nitro, show similar trends but with different and less well documented substituent effects.

1.19 Polycyclic aromatic hydrocarbons

The range of polycyclic aromatic hydrocarbons is too great for detailed consideration in this book. Because there are many energy levels between which electronic transitions can take place, their spectra are usually complicated, and for that reason are useful as fingerprints. When they only have relatively non-polar substituents, such as alkyl and acetoxy groups, the spectra are similar in the shape and position of the absorption peaks to the unsubstituted hydrocarbons. The degradation products of natural products often contain polycyclic nuclei which can be identified in this way as, for example, a phenanthrene or a perylene. The spectra of a typical series, naphthalene, anthracene, and naphthacene, are illustrated in Fig. 1.8; note that this figure uses a logarithmic ordinate in order to encompass the range of intensities.

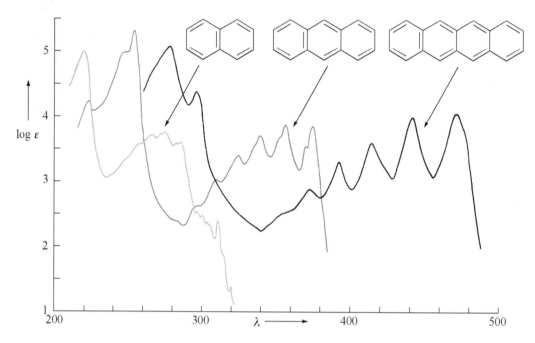

Fig. 1.8
(Spectra taken with permission from R. A. Friedel and M. Orchin, *Ultraviolet Spectra of Aromatic Compounds*, Wiley, New York, 1951.)

1.20 Heteroaromatic compounds

In general heteroaromatic compounds resemble the spectra of their corresponding hydrocarbons, but only in the crudest way. The heteroatom, whether like that in a pyrrole or that in a pyridine, leads to pronounced substituent effects which depend on the electron-donating or withdrawing effect of the substituent and the heteroatom and on their orientation. The effects of these factors are predictable in a qualitative way using the same sorts of criteria as were used in Sec. 1.18 when considering the effects of more than one substituent on a benzene ring. For example, a simple pyrrole **15** and a pyrrole with an electron-withdrawing substituent **16** have strikingly different absorption maxima. The conjugation present from the nitrogen lone pair through the pyrrole ring to the carbonyl group increases the length of the chromophore and leads to longer wavelength absorption. The conjugation to the 5-position in the pyrrole **16** provides a longer conjugated system than the conjugation to the 3-position in the pyrrole **17**, giving rise to a longer wavelength and more intense absorption. The following illustrations of some common heterocyclic systems, including the four nucleoside bases **23-26**, give some indication of the spectra observed.

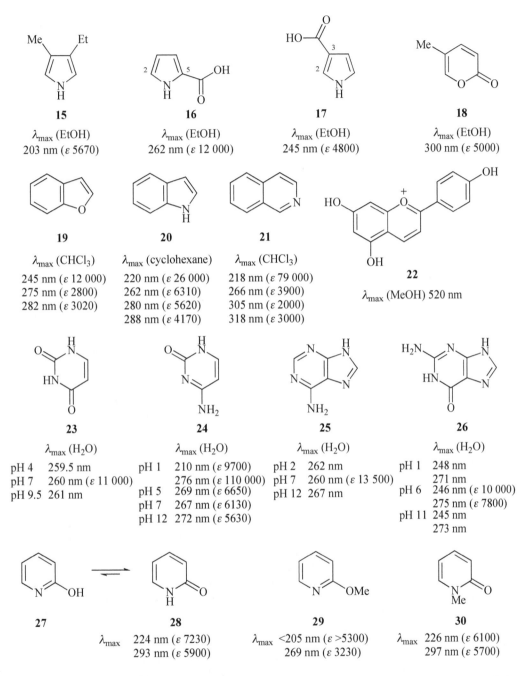

15

λ_max (EtOH)
203 nm (ε 5670)

16

λ_max (EtOH)
262 nm (ε 12 000)

17

λ_max (EtOH)
245 nm (ε 4800)

18

λ_max (EtOH)
300 nm (ε 5000)

19

λ_max (CHCl₃)

245 nm (ε 12 000)
275 nm (ε 2800)
282 nm (ε 3020)

20

λ_max (cyclohexane)

220 nm (ε 26 000)
262 nm (ε 6310)
280 nm (ε 5620)
288 nm (ε 4170)

21

λ_max (CHCl₃)

218 nm (ε 79 000)
266 nm (ε 3900)
305 nm (ε 2000)
318 nm (ε 3000)

22

λ_max (MeOH) 520 nm

23

λ_max (H₂O)

pH 4   259.5 nm
pH 7   260 nm (ε 11 000)
pH 9.5 261 nm

24

λ_max (H₂O)

pH 1   210 nm (ε 9700)
       276 nm (ε 110 000)
pH 5   269 nm (ε 6650)
pH 7   267 nm (ε 6130)
pH 12 272 nm (ε 5630)

25

λ_max (H₂O)

pH 2   262 nm
pH 7   260 nm (ε 13 500)
pH 12 267 nm

26

λ_max (H₂O)

pH 1   248 nm
       271 nm
pH 6   246 nm (ε 10 000)
       275 nm (ε 7800)
pH 11 245 nm
       273 nm

27   **28**

λ_max   224 nm (ε 7230)
        293 nm (ε 5900)

29

λ_max   <205 nm (ε >5300)
        269 nm (ε 3230)

30

λ_max   226 nm (ε 6100)
        297 nm (ε 5700)

In the case of tautomeric molecules it is sometimes possible, using ultraviolet spectroscopy, to identify which is the stable tautomer. For example, the equilibrium between 2-hydroxypyridine **27** and pyrid-2-one **28** has been shown to lie far to the right; the ultraviolet spectrum of the solution resembles that of a solution of *N*-methylpyrid-2-one **30** and is different from that of 2-methoxypyridine **29**. The change in the absorption maxima with the change of pH in tautomeric molecules is due sometimes to a change in the chromophore as a result of the tautomerism and sometimes to simple protonation or deprotonation. This point is mentioned here in order to stress the importance of careful control of the medium in which spectra are taken. The changes in absorption maxima with change of pH are useful diagnostically, since they serve in some systems to identify the pattern of substitution.

1.21 Quinones

31	**32**	**33**	**34**
λ_{max}	λ_{max}	λ_{max}	λ_{max}
(hexane)	(EtOH)	(hexane)	(EtOH)
242 nm (ε 24 000)	276 nm (ε 2000)	241 nm (ε 20 000)	243.5 nm (ε 33 000)
281 nm (ε 400)	387 nm (ε 800)	246 nm (ε 23 500)	252.5 nm (ε 51 500)
434 nm (ε 20)		251 nm (ε 19 000)	263 nm (ε 20 000)
		256 nm (ε 13 000)	272 nm (ε 20 000)
		330 nm (ε 2750)	325 nm (ε 5600)
			405 nm (ε 90)

The quinones **31-34** are a representative series of these usually coloured compounds. The colour of the simpler members is from the weak n→π* transition, similar to that of α-diketones.

1.22 Corroles, chlorins and porphyrins

Fig. 1.9 shows the visible spectra of representative members of each of the three main classes of pyrrole pigments: hydrogenobyrinic acid **35**, with the chromophore of vitamin B$_{12}$, chlorophyll **36**, and protoporphyrin IX **37**. The long conjugated systems, with the chromophores emphasised in black, give rise to an exceptionally strong and sharp band, the shoulder of which can be seen on the left in each spectrum in Fig. 1.9. In chlorins and porphyrins this band is called the Soret band, and it occurs near 400 nm (ε 100 000). Changes in the chromophore in each class can often be recognised by changes in the position and relative intensity of the four or more weaker, but still strong, bands found in the visible region. These spectra illustrate the general principle that the longer conjugated systems lead to more intense absorption at longer wavelength, and they also reveal how the detailed pattern changes diagnostically with the presence or absence of extra double bonds. The chlorophyll spectrum, in particular, shows the strong absorption at the blue and the red end of the spectrum, leaving the green colour with which we are so familiar. Another conjugated macrocyclic aromatic system, [18]-annulene, shows a similar intense band at 369 nm (ε 303 000).

The pyrrole pigments are mentioned here to stress the importance and usefulness of ultraviolet and visible spectroscopy in the study of groups of compounds possessing a long, complicated chromophore. The large number of model systems available makes it

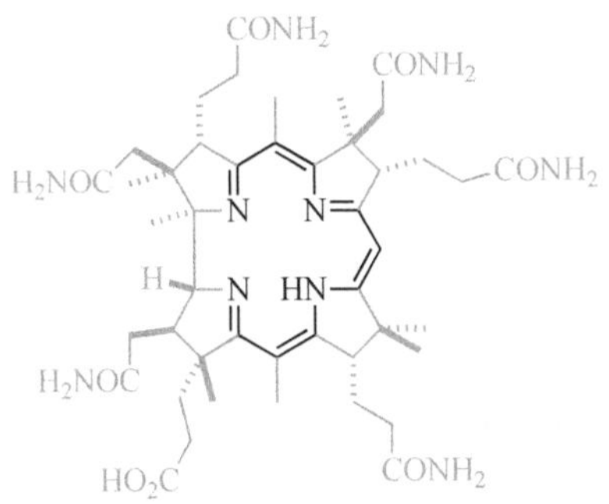

35 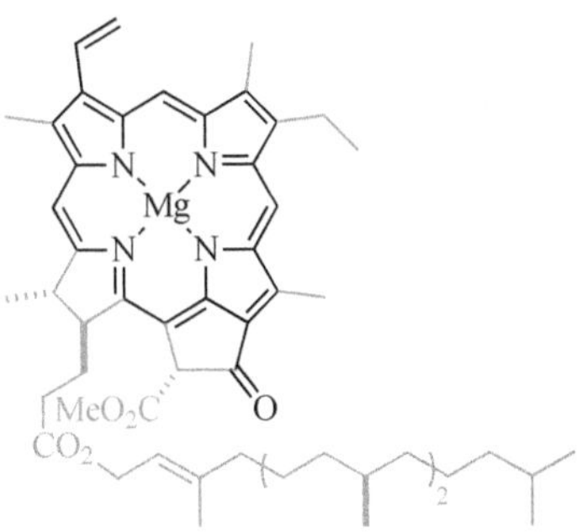   **36** 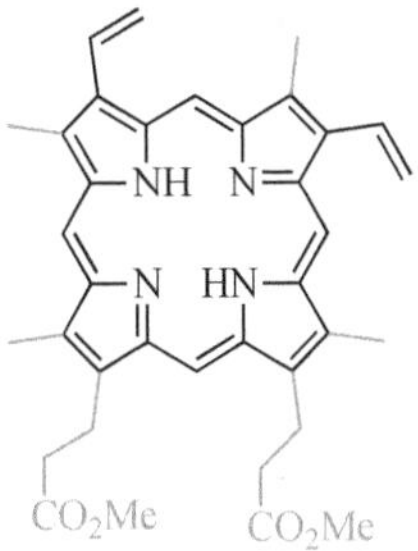   **37**

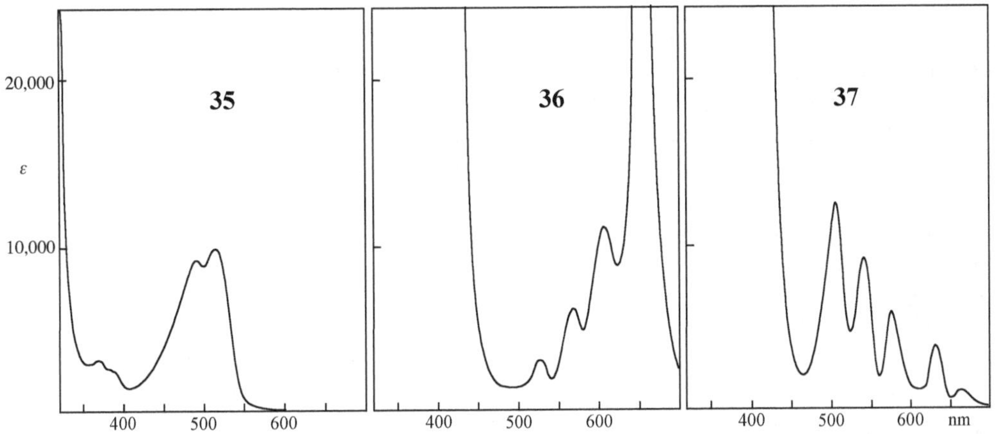

Fig. 1.9

relatively easy to recognise a chromophore, which the other spectroscopic methods do not probe. For example, the oxidations involved in the biodegradation of chlorophyll and haem interrupt the conjugated system in the middle, and are immediately picked up by the very dramatic changes in the visible absorption spectra.

1.23 Non-conjugated interacting chromophores

Non-conjugated systems usually have little effect on each other; diphenyl methane has a spectrum similar to that of toluene; the cross-conjugation of the trienone **10** was successfully ignored when calculating the expected absorption maximum; and even diphenyl ether is not very different from anisole. However, conjugation through space is possible when an auxochrome is held close enough to a conjugated system without being directly conjugated to it. Thus, the $\beta\gamma$-unsaturated ketone norbornenone **38** shows n$\rightarrow\pi^*$ and $\pi\rightarrow\pi^*$ transitions shifted towards the blue relative to the absorptions of the isolated components. There is evidently some conjugation across space raising the HOMO and lowering the LUMO, but not as effectively as in $\alpha\beta$-unsaturated ketones. Also, the cumulated double bonds in allenes **39** and ketenes **40**, although not formally conjugated, cause some ultraviolet light to be absorbed weakly in the accessible region.

	38	**39**	**40**
λ_{max}	210 nm (ε 3000)	170 nm (ε 4000)	227 nm (ε 360)
	305 nm (ε 290)	227 nm (ε 630)	375 nm (ε 20)

1.24 The effect of steric hindrance to coplanarity

Steric hindrance preventing full coplanarity in a conjugated system, as in *cis*-stilbene **41**, interferes with the effect conjugation has on the energy of the HOMO and the LUMO. In a poorly conjugated system, the HOMO is not raised in energy to the same extent, and the LUMO is not lowered in energy to the same extent. As a result the gap (z in Fig. 1.2) is larger. The result for the longest wavelength absorption band in *cis*-stilbene **41** is a shorter wavelength and less intense maximum than for *trans*-stilbene **42**. Similarly, the *ortho* substituents in 2,4,6-trimethylacetophenone **43** (R = Me) prevent the carbonyl group from lying coplanar with the benzene ring; this ketone has weaker absorption at shorter wavelength than *p*-methylacetophenone **43** (R = H).

	41	**42**	**43**	**44**
			R = Me	R = Me
λ_{max}	224 nm (ε 24 400)	228 nm (ε 16 400)	λ_{max} 242 nm (ε 3200)	λ_{max} 385 nm (ε 4840)
(EtOH)	280 nm (ε 10 500)	295.5 nm (ε 29 000)	R = H	R = H
			λ_{max} 252 nm (ε 15 000)	λ_{max} 375 nm (ε 16 000)

On the other hand, the absorption maximum of 3,5-dimethyl-*p*-nitroaniline **44** (R = Me) shows the usual reduction in intensity but this time a red shift relative to that of the parent compound *p*-nitroaniline **44** (R = H). It is possible that the former absorption is from a different transition to that monitored in the latter case.

45	**46**
no strong absorption >210 nm	λ_{max} 227 nm (ε 5500)

The dilactone **45** produced from shellolic acid misleadingly showed no maximum in the accessible ultraviolet region, but hydrolysis gave a product which showed the expected absorption for an αβ-unsaturated acid **46**. The steric constraints in the polycyclic structure had prevented effective conjugation between the double bond and the carbonyl group, but the release of this constraint allowed the two π-bonds to overlap.

This observation provides an opportunity to stress that changes between the ultraviolet spectrum of a starting material and a product make it one of the easiest tools to use for following the kinetics of a chemical reaction, and that ultraviolet spectroscopy is possibly used more for this purpose than in structure determination. Nevertheless, the immediate and highly sensitive detection of conjugated systems is still a powerful application for this the oldest of the spectroscopic methods.

1.25 Internet

The Internet is a continuously evolving system, with links and protocols changing frequently. The following information is inevitably incomplete and may no longer apply, but it gives you a guide to what you can expect. Some websites require particular operating systems and may only work with a limited range of browsers, some require payment, and some require you to register and to download programs before you can use them.

For guides to spectroscopic data on the Internet, see the websites at MIT, the University of Waterloo and the University of Texas, representative of several others. They are tailored for internal use, but are informative nevertheless:

http://libraries.mit.edu/guides/subjects/chemistry/spectra_resources.htm
http://lib.uwaterloo.ca/discipline/chem/spectral_data.html
http://www.lib.utexas.edu/chem/info/spectra.html

Ultraviolet spectroscopy is not as well served on the Internet as the other spectroscopic methods. The set of books, *Organic Electronic Spectral Data*, is still the best source for UV and visible spectroscopic data.

There is a database of 1600 compounds with UV data on the NIST website belonging to the United States Secretary of Commerce:

http://webbook.nist.gov/chemistry/name-ser.html

Type in the name of the compound you want, check the box for UV/Vis spectrum, and click on Search, and if the ultraviolet spectrum is available it will show it to you.

ACD (Advanced Chemistry Development) Spectroscopy sell proprietary software called ACD/SpecManager that handles all four spectroscopic methods, as well as other analytical tools:

http://www.acdlabs.com/products/spec_lab/exp_spectra/

It is able to process and store the output of the instruments that take spectra, and can be used to catalogue, share and present your own data. It also gives access to a few free databases for UV spectra.

Wiley-VCH keep an up-to-date website on their spectroscopic books and provide links. The URL giving access to information about spectroscopy, including UV, is:

http://www.spectroscopynow.com/Spy/basehtml/SpyH/1,1181,7-4-773-0-773-directories--0,00.html

1.26 Bibliography

DATA

Organic Electronic Spectral Data, Wiley, New York, Vols. 1-31 (1960-96).

Sadtler Handbook of Ultraviolet Spectra, Sadtler Research Laboratories, 1979.

D. M. Kirschenbaum, ed., *Atlas of Protein Spectra in the Ultraviolet and Visible Region*, Plenum, New York, 1972.

TEXTBOOKS

H. H. Jaffe and M. Orchin, *Theory and Applications of Ultraviolet Spectroscopy*, Wiley, New York, 1962.

J. N. Murrell, *The Theory of the Electronic Spectra of Organic Molecules*, Methuen, London, 1963.

G. R. Barrow, *Introduction to Molecular Spectroscopy*, McGraw-Hill, New York, 1964.

E. F. H. Brittain, W. O. George and C. H. J. Wells, *Introduction to Molecular Spectroscopy*, Academic Press, London, 1970.

A. I. Scott, *Interpretation of the Ultraviolet Spectra of Natural Products*, Pergamon Press, Oxford, 1964.

S. F. Mason, Chapter 7, The Electronic Absorption Spectra of Heterocyclic Compounds, in *Physical Methods in Heterocyclic Chemistry*, Vol. II, Academic Press, New York, 1963.

C. N. R. Rao, *Ultraviolet and Visible Spectroscopy*, Butterworths, London, 3rd Ed., 1975.

M. J. K. Thomas, *Ultraviolet and Visible Spectroscopy*, J. Wiley, Chichester, 2nd Ed., 1996.

2. Infrared spectra

2.1 Introduction

The infrared spectra of organic compounds are associated with transitions between vibrational energy levels. Molecular vibrations may be detected and measured either in an infrared spectrum or indirectly in a Raman spectrum. The most useful vibrations, from the point of view of the organic chemist, occur in the narrower range of 2.5-16 μm (1 μm = 10^{-6} m). The position of an absorption band in the spectrum may be expressed in microns (μm), but standard practice uses a frequency scale in the form of wavenumbers, which are the reciprocals of the wavelength, cm^{-1}. The useful range of the infrared for an organic chemist is between 4000 cm^{-1} at the high-frequency end and 625 cm^{-1} at the low-frequency end.

Many functional groups have vibration frequencies, characteristic of that functional group, within well-defined regions of this range; these are summarised in Charts 1-4 at the end of this chapter, with more detail in the tables of data that follow. Because many functional groups can be identified by their characteristic vibration frequencies, the infrared spectrum is the simplest, most rapid, and often most reliable means for identifying the functional groups.

Equation 2.1, which is derived from the model of a mass m vibrating at a frequency v on the end of a fixed spring, is useful in understanding the range of values of the vibrational frequencies of various kinds of bonds.

$$v = \sqrt{\frac{k}{m}} \tag{2.1}$$

where k is a measure of the strength of the spring. However, in chemical bonds, one end of the 'spring' (bond) is not fixed, but rather there are two masses (m_1 and m_2) involved and each is able to move. The m of Eq. 2.1 is now determined by the relationship in Eq. 2.2.

$$\frac{1}{m} = \frac{1}{m_1} + \frac{1}{m_2} \tag{2.2}$$

If one of the masses (say, m_1) is infinitely large, $1/m_1$ is then zero, and the relevant mass m for Eq. 2.1 is simply that of m_2—making it analogous to the case where one end of the 'spring' is fixed.

Simple substitutions of masses in these equations allow us to understand that, with other things being equal: (i) C—H bonds will have higher stretching frequencies than C—C bonds, which in turn are likely to be higher than C—halogen bonds; (ii) O—H bonds will have higher stretching frequencies than O—D bonds; and (iii), since k

27

increases with increasing bond order, the relative stretching frequencies of carbon-carbon bonds lie in the order: $C\equiv C > C=C > C—C$.

These generalisations are useful, and Eqs. 2.1 and 2.2 allow an increased understanding of the empirical data that are subsequently presented in this chapter. You may often be able to extend the use of the model in a way that will make it easier to understand the trends that are observed. However, because of the other variables that influence vibrational frequencies, the equations should be taken as no more than a frequently useful guide.

2.2 Preparation of samples and examination in an infrared spectrometer

Older spectrometers used a source of infrared light which had been split into two beams of equal intensity. Only one of these was passed through the sample, and the difference in intensities of the two beams was then plotted as a function of wavenumber. Using this old technology, a scan typically took about 10 minutes. Most spectrometers in use today use a Fourier transform method, and the spectra are called Fourier transform infrared (FTIR) spectra. A source of infrared light, emitting radiation throughout the whole frequency range of the instrument, typically 4600-400 cm^{-1}, is again divided into two beams of equal intensity. Either one beam is passed through the sample, or both are passed, but one beam is made to traverse a longer path than the other. Recombination of the two beams produces an interference pattern that is the sum of all the interference patterns created by each wavelength in the beam. By systematically changing the difference in the two paths, the interference patterns change to produce a detected signal varying with optical path difference, as modified by the selective absorption by the sample of some frequencies. This pattern is known as the interferogram, and looks nothing like a spectrum. However, Fourier transformation of the interferogram, using a computer built into the instrument, converts it into a plot of absorption against wavenumber just like that from the older method. There are several advantages to FTIR over the older method, and few disadvantages. Because it is not necessary to scan each wavenumber successively, the whole spectrum is measured in at most a few seconds. Because it is not dependent upon a slit and a prism or grating, high resolution in FTIR is easier to obtain without sacrificing sensitivity. FTIR is especially useful for examining small samples (several scans can be added together) and for taking the spectrum of compounds produced only for a short period in the outflow of a chromatograph. Finally, the digital form in which the data are handled in the computer allows for adjustment and refinement. For example, by subtracting the background absorption of the medium in which the spectrum was taken, or by subtracting the spectrum of a known impurity from that of a mixture to reveal the spectrum of the pure component. However, the way in which infrared spectra are taken does not affect their appearance. The older spectra and FTIR spectra look very similar, and older spectra in the literature are still valuable for comparison. Compounds may be examined in the vapour phase, as pure liquids, in solution, and in the solid state.

In the vapour phase. The vapour is introduced into a cell, usually about 10 cm long, which can then be placed directly in the path of one of the infrared beams. The end walls of the cell are usually made of sodium chloride, which is transparent to infrared in the usual range. Most organic compounds have too low a vapour pressure for this phase to be useful.

As a liquid. A drop of the liquid is squeezed between flat plates of sodium chloride (transparent through the 4000-625 cm^{-1} region). This is the simplest of all procedures.

Alternatively, if the sample of the liquid is not suitable for dispensing as a drop, a solution in a volatile and dry solvent may be deposited directly onto the surface of a sodium chloride plate, and the solvent allowed to evaporate in a dry atmosphere to leave a thin film.

In solution. The compound is dissolved to give, typically, a 1-5% solution in carbon tetrachloride or, for its better solvent properties, alcohol-free chloroform. This solution is introduced into a cell, 0.1-1 mm thick, made of sodium chloride. A second cell of equal thickness, but containing pure solvent, is placed in the path of the other beam of the spectrometer in order that solvent absorptions should be balanced. Spectra taken in such dilute solutions in non-polar solvents are generally the most desirable, because they are normally better resolved than spectra taken on solids, and also because intermolecular forces, which are especially strong in the crystalline state, are minimised. On the other hand, many compounds are not soluble in non-polar solvents, and all solvents absorb in the infrared; when the solvent absorption exceeds about 65% of the incident light, useful spectra cannot be obtained because insufficient light is transmitted to work the detection mechanism efficiently. Carbon tetrachloride and chloroform, fortunately, absorb over 65% of the incident light only in those regions (Fig. 2.1) which are of little interest in diagnosis. Other solvents, of course, may be used but the areas of usefulness in each case should be checked beforehand, taking account of the size of the cell being used. In rare cases aqueous solvents are useful; special calcium fluoride cells are then used.

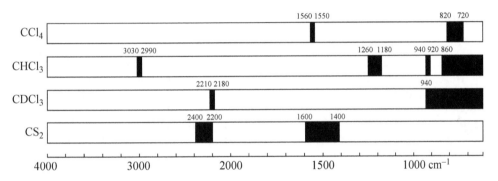

Fig. 2.1 Darkened areas are the regions in which the solvent cannot be used with a 0.2 mm cell

In the solid state. About 1 mg of a solid is finely ground in a small agate mortar with a drop of a liquid hydrocarbon (Nujol, Kaydol) or, if C—H vibrations are to be examined, with hexachlorobutadiene. The mull is then pressed between highly polished flat plates of sodium chloride. Alternatively, the solid, often much less than 1 mg, is ground with 10-100 times its bulk of pure potassium bromide and the mixture pressed into a disc using a mould and a hydraulic press. The use of KBr eliminates the problem (usually not troublesome) of bands from the mulling agent and tends, on the whole, to give rather better spectra, except that a band at 3450 cm^{-1}, from the O—H group of traces of water, almost always appears (see Fig. 2.7). Solids may also be deposited, either from a melt or, as with liquids described above, by evaporation from a solution directly onto the surface of a sodium chloride plate, with a sacrifice, usually small, from scattering off a crystalline surface. Because of intermolecular interactions, band positions in solid state spectra are often different from those of the corresponding solution spectra. This is particularly true of those functional groups which take part in hydrogen bonding. On the other hand, the number of resolved lines is often greater in solid state spectra, so that comparison of the spectra of, for example, synthetic and natural samples in order to determine identity is best done in the solid state. This is only true, of course, when the same crystalline modification is in use; racemic, synthetic material, for example, should be compared with enantiomerically pure, natural material in solution.

2.3 Examination in a Raman spectrometer

Raman spectra are generally taken on instruments using laser sources, and the quantity of material needed is now of the order of a few mg. A liquid or a concentrated solution is irradiated with monochromatic light, and the scattered light is examined by a spectrometer using photoelectric detection. Most of the scattered light consists of the parent line produced by absorption and re-emission. Much weaker lines, which constitute the Raman spectrum, occur at lower and higher energy and are caused by absorption and re-emission of light coupled with vibrational excitation or decay, respectively. The difference in frequency between the parent line and the Raman line is the frequency of the corresponding vibration. Raman spectroscopy is not used by organic chemists routinely for structure determination, but for the detection of certain functional groups (see Fig. 2.12) and for the analysis of mixtures—of deuterated compounds for example—it has found some use, especially by analytical chemists.

2.4 Selection rules

If the frequency of a vibration of the sample molecule falls within the range of the instrument, the molecule may absorb energy of this frequency from the light, but only when the oscillating dipole moment (from a molecular vibration) interacts with the oscillating electric vector of the infrared beam. A simple rule for deciding if this interaction (and hence absorption of light) occurs is that the dipole moment at one extreme of a vibration must be different from the dipole moment at the other extreme of the vibration. In the Raman effect a corresponding interaction occurs between the light and the molecule's polarisability, resulting in different selection rules.

The most important consequence of these selection rules is that in a molecule with a centre of symmetry those vibrations symmetrical about the centre of symmetry are active in the Raman and inactive in the infrared (see Fig. 2.12); those vibrations which are not centrosymmetric are inactive in the Raman and usually active in the infrared. This is doubly useful, for it means that the two types of spectrum are complementary. Furthermore, the more easily obtained, the infrared, is the more useful, because most functional groups are not centrosymmetric. The symmetry properties of a molecule in a solid can be different from those of an isolated molecule. This can lead to the appearance of infrared absorption bands in a solid state spectrum which would be forbidden in solution or in the vapour phase.

2.5 The infrared spectrum

A complex molecule has many vibrational modes which involve the whole molecule. To a good approximation, however, many of these molecular vibrations are largely associated with the vibrations of individual bonds and are called localised vibrations. These localised vibrations are useful for the identification of functional groups, especially the stretching vibrations of O—H and N—H single bonds and all kinds of triple and double bonds, almost all of which occur with frequencies greater than 1500 cm^{-1}. The stretching vibrations of other single bonds, most bending vibrations and the soggier vibrations of the molecule as a whole give rise to a series of absorption bands at lower energy, below 1500 cm^{-1}, the positions of which are characteristic of that molecule. The net result is a region above 1500 cm^{-1} showing absorption bands assignable to a number of functional groups, and a region containing many bands, characteristic of the compound in question and no other compound, below 1500 cm^{-1}. For obvious reasons, this is called the fingerprint region.

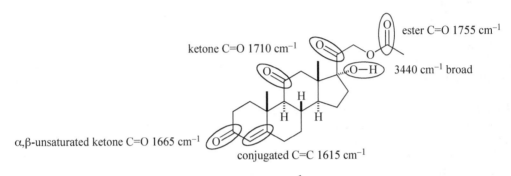

1

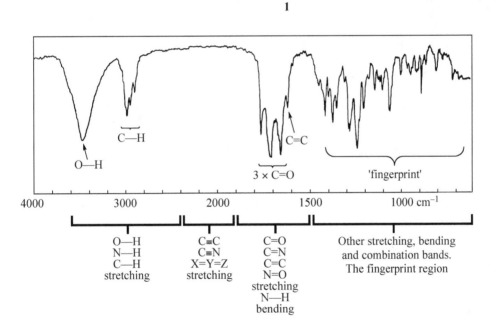

Fig. 2.2 The infrared spectrum of cortisone acetate **1**

Fig. 2.2 shows a representative infrared spectrum, that of cortisone acetate **1**. It shows the strong absorption from the stretching vibrations above 1500 cm^{-1} demonstrating the presence of each of the functional groups: the O—H group, three different C=O groups and the weaker absorption of the C=C double bond, as well as displaying a characteristic fingerprint below 1500 cm^{-1}.

By convention absorbance is plotted downwards, opposite to the convention for ultraviolet spectra, but the maxima are still called peaks or bands. Rotational fine structure is smoothed out, and the intensity is frequently not recorded. When intensity is recorded, it is usually expressed subjectively as strong (s), medium (m) or weak (w). To obtain a high-quality spectrum, the quantity of substance is adjusted so that the strongest peaks absorb something close to 90% of the light. The scale on the abscissa is linear in frequency, but most instruments change the scale, either at 2200 cm^{-1} or at 2000 cm^{-1} to double the scale at the low-frequency end. The ordinate is linear in percent transmittance, with 100% at the top and 0% at the bottom.

The regions in which the different functional groups absorb are summarised below Fig. 2.2. The stretching vibrations of single bonds to hydrogen give rise to the absorption at the high-frequency end of the spectrum as a result of the low mass of the hydrogen atom, making it easy to detect the presence of O—H and N—H groups. Since most organic compounds have C—H bonds, the absorption close to 3000 cm^{-1} is rarely useful, although C—H bonds attached to double and triple bonds can be usefully identified. Thereafter, the order of stretching frequencies follows the order: triple bonds at higher frequency than double bonds and double bonds higher than single bonds—on the whole the greater the strength of the bond between two similar atoms the higher the frequency of the vibration. Bending vibrations are of lower frequency and usually appear in the fingerprint region below 1500 cm^{-1}, but one exception is the N—H bending vibration, which appears in the 1600-1500 cm^{-1} region. Polystyrene is sometimes used to provide accurately placed calibration lines at 2924, 1603, 1028 and 906 cm^{-1}.

Although many absorption bands are associated with the vibrations of individual bonds, other vibrations are coupled vibrations of two or more components of the whole molecule. Whether localised or not, stretching vibrations are given the symbol ν, and the various bending vibrations are given the symbol δ. Coupled vibrations may be subdivided into asymmetric and symmetric stretching, and the various bending modes into scissoring, rocking, wagging and twisting, as defined for a methylene group in Fig. 2.3. A coupled asymmetric and symmetric stretching pair is also found with many other groups, like carboxylic anhydrides, carboxylate ions and nitro groups, each of which has two equal bonds close together.

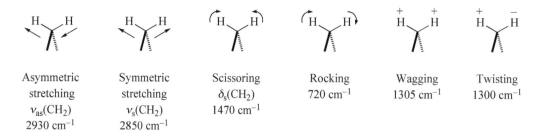

| Asymmetric stretching $\nu_{as}(CH_2)$ 2930 cm^{-1} | Symmetric stretching $\nu_s(CH_2)$ 2850 cm^{-1} | Scissoring $\delta_s(CH_2)$ 1470 cm^{-1} | Rocking 720 cm^{-1} | Wagging 1305 cm^{-1} | Twisting 1300 cm^{-1} |

Fig. 2.3 Localised vibrations of a methylene group

2.6 The use of the tables of characteristic group frequencies

Reference charts and tables of data are collected together at the end of this chapter for ready reference. Each of the three frequency ranges above 1500 cm^{-1} shown in Fig. 2.2 is expanded to give more detail in Charts 1-4 in Sec. 2.13. These charts summarise the narrower ranges within which each of the functional groups absorbs. The absorption bands which are found in the fingerprint region and which are assignable to functional groups are occasionally useful, either because they are sometimes strong bands in otherwise featureless regions or because their absence may rule out incorrect structures, but such identifications should be regarded as helpful rather than as definitive, since there are usually many bands in this area. Tables of detailed information can be found in Sec. 2.14 at the end of this chapter, arranged by functional groups roughly in descending order of their stretching frequencies.

One could deal with the spectrum of an unknown as follows. Examine each of the three main regions of the spectrum above the fingerprint region, as identified on Fig. 2.2; at this stage certain combinations of structures can be ruled out—the absence of O—H or C=O, for example—and some tentative conclusions reached. Where there is still ambiguity—which kind of carbonyl group, for example—the tables corresponding to those groups that might be present should be consulted. It is important to be sure that the bands under consideration are of the appropriate intensity for the structure suspected. A *weak* signal in the carbonyl region, for example, is not good evidence that a carbonyl group is present, since carbonyl stretching is always strong—it is more likely to be an overtone or to have been produced by an impurity.

The text following this section amplifies some of the detail for each of the main functional groups, and shows the appearance, sometimes characteristic, of several of the bands. Cross-reference to the tables at the end is inevitable and will need to be frequent.

2.7 Absorption frequencies of single bonds to hydrogen 3600-2000 cm^{-1}

C—H Bonds. The precise position of the various CH, CH_2, and CH_3 symmetrical and unsymmetrical vibration frequencies are well known. C—H bonds do not take part in hydrogen bonding and so their position is little affected by the state of measurement or their chemical environment. C—C vibrations, which absorb in the fingerprint region, are generally weak and not practically useful. Since many organic molecules possess saturated C—H bonds, their absorption bands, stretching in the 3000-2800 cm^{-1} region and bending in the fingerprint region, are of little diagnostic value, but a few special structural features in saturated C—H groupings do give rise to characteristic absorption bands (Table 2.1). Thus, methyl and methylene groups usually show two sharp bands from the symmetric and asymmetric stretching (Fig. 2.3), which can sometimes be picked out, but the general appearance of the accumulation of all the saturated C—H stretching vibrations often leads to broader and not fully resolved bands like those illustrated in many of the spectra below. The absence of saturated C—H absorption in a spectrum is, of course, diagnostic evidence for the absence of such a part structure in the corresponding compound. Unsaturated and aromatic C—H stretching frequencies (Table 2.1) can be distinguished from the saturated C—H absorption, since they occur a little above 3000 cm^{-1} and are relatively weak, as in the spectrum of ethyl benzoate 2 (Fig. 2.4) and benzonitrile 14 (Fig. 2.7). Terminal acetylenes give rise to a characteristic strong, sharp line close to 3300 cm^{-1} from ≡C—H stretching, as in the spectrum of hexyne 3 (Fig. 2.4). A C—H bond antiperiplanar to a lone pair on oxygen or nitrogen is weakened, and the stretching frequency is lowered. Thus, the C—H bond of aldehydes gives rise to a relatively sharp band close to 2760 cm^{-1}, as seen in the spectrum of heptanal 4 (Fig. 2.4), and ethers and amines also show bands in the low-frequency region 2850-2750 cm^{-1}. When the antiperiplanar arrangement is rigidly fixed, as in axially-oriented C—H bonds in six-membered cyclic amines, C—H stretching has an unusually low frequency, giving rise to absorption known as Bohlmann bands.

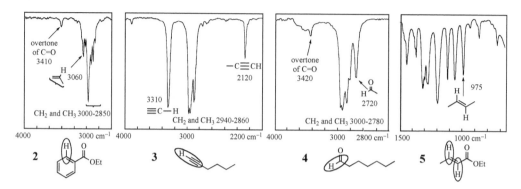

Fig. 2.4 Some characteristic C—H absorptions in the infrared

The C—H bending vibrations are in the fingerprint region, with methine C—H bending and CH_3 and CH_2 symmetrical bending giving rise in many organic compounds to two bands close to 1450 and 1380 cm^{-1}, as seen in the common mulling agent Nujol. The out-of-plane vibration of *trans* -CH=CH- double bonds is one of the more usefully diagnostic bending vibrations. It occurs in a narrow range 970-960 cm^{-1}, or at slightly higher frequency if conjugated, and it is always strong. In contrast, the corresponding vibration of the *cis* isomer is of lower intensity and at lower frequency, typically in the range 730-675 cm^{-1}. The band at 975 cm^{-1} in the fingerprint of ethyl *trans*-crotonate **5** (Fig. 2.4) clearly shows that such a feature may be present; if it were not there, it would be diagnostic of the *absence* of this feature, as in the spectrum of the *cis*-alkene **20** in Fig. 2.12.

O—H Bonds. The value of the O—H stretching frequency (Table 2.3) has been used for many years as a test for, and measure of, the strength of hydrogen bonds. The stronger the hydrogen bond the longer the O—H bond, the lower the vibration frequency, and the broader and the more intense the absorption band. O—H bonds not involved in hydrogen bonding have a sharp band in the 3650-3590 cm^{-1} range, typically observed for samples in the vapour phase, in very dilute solution, or when such factors as steric hindrance prevent hydrogen bonding. Frequently solution phase spectra show both bands, as seen in the somewhat hindered alcohol **6** (Fig. 2.5). Pure liquids, solids, and many solutions, on the other hand, show only the broad strong band in the 3600-3200 cm^{-1} range, because of exchange and because of the different degrees of hydrogen bonding present within the sample, as seen in the spectrum of the alcohol **7** (Fig. 2.5) taken with a neat sample.

Weak intramolecular hydrogen bonds, like those in 1,2-diols for example, show a sharp band in the range 3570-3450 cm^{-1}, the precise position being a measure of the strength of the hydrogen bond. Strong intramolecular hydrogen bonds usually give rise to broad and strong absorption in the 3200-2500 cm^{-1} range. When a carbonyl group is the hydrogen bond acceptor, its characteristic stretching frequency is lowered, as seen for example in the dimeric association of most carboxylic acids. A broad absorption in the 3200-2500 cm^{-1} range, usually seen under and surrounding the C—H absorption, accompanied by carbonyl absorption in the 1710-1650 cm^{-1} region, is highly

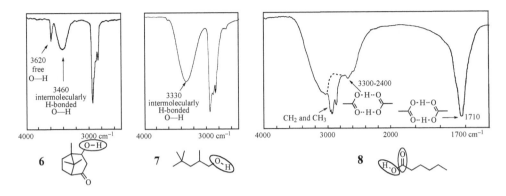

Fig. 2.5 Some characteristic O—H absorptions in the infrared

characteristic of carboxylic acids, like hexanoic acid **8** (Fig. 2.5), or of their vinylogues, β-dicarbonyl compounds in the enol form. Distinctions can be made among the various hydrogen-bonding possibilities by testing the effect of dilution: intramolecular hydrogen bonds are unaffected, while intermolecular hydrogen bonds are broken, leading to an increase in—or the appearance of—free O—H absorption. Spectra taken of samples in the solid state almost always show only the broad strong band in the range 3400-3200 cm^{-1}. Replacing an H with a D atom leads to absorption at a frequency 0.73 times lower. Note that the weak bands in the O—H region of the spectra of the ester **2** and the aldehyde **4** are clearly not O—H stretching, because they are too weak. S—H bond stretching is significantly lower in frequency, typically appearing as a weak and slightly broadened band near 2600 cm^{-1}.

N—H Bonds. The stretching frequencies of the N—H bonds of amines (Table 2.4) are typically in the range 3500-3300 cm^{-1}. They are less intense than those of O—H bonds but can easily be confused with hydrogen-bonded O—H stretching. Because an N—H has a weaker tendency to form a hydrogen bond, its absorption is often sharper, and can be very sharp, as in the N—H band shown by the indole N—H from tryptophan **9** (Fig. 2.6). Even in very dilute solutions N—H bonds never give rise to absorption as high as the free O—H near 3600 cm^{-1}. Primary amines and amides like the amide **10** (Fig. 2.6) give rise to two bands, typically one at 3500 and the other at 3400 cm^{-1}, because there are

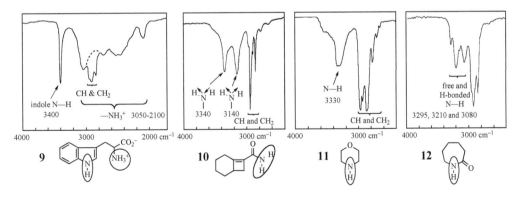

Fig. 2.6 Some characteristic N—H absorptions in the infrared

two stretching vibrations, one symmetrical and the other unsymmetrical, similar to those of a methylene group (Fig. 2.3). Secondary amines like morpholine **11** only have the one band. Secondary amides in an s-*trans* configuration have only one band, but lactams, in which the amide group is in an s-*cis* configuration, often show several bands from various hydrogen-bonded associations, especially in the solid state, as in the spectrum of caprolactam **12** (Fig. 2.6).

Amine salts and the zwitterions of amino acids give rise to several N—H stretching bands on the low-frequency side of any C—H absorption, and sometimes reaching as low as 2000 cm^{-1}, as in the spectrum of tryptophan **9** (Fig. 2.6).

The N—H bending vibration of primary and secondary amides appears just above the fingerprint region, and is responsible for what is called the amide-II band, which is discussed in the section on the carbonyl group.

2.8 Absorption frequencies of triple and cumulated double bonds 2300-1930 cm^{-1}

Terminal acetylenes absorb in the narrow range 2140-2100 cm^{-1}, as in the spectrum of hexyne **3** in Fig. 2.4. Internal acetylenes absorb in the range 2260-2150 cm^{-1}, with conjugated triple bonds and enynes at the lower end of the range (Table 2.6). Note that the differences observed between the range of absorption frequencies for terminal and internal acetylenes is consistent with the expectations on the bases of Eqs. 2.1 and 2.2. In the case of an internal acetylene, the acetylenic carbons will both behave as though they have somewhat larger masses than will the carbon of an acetylenic C—H group. The consequence is that internal acetylenes should absorb at the higher frequency—as is indeed observed. The absorptions are often weak, unless they are conjugated, because of the small change in dipole moment on stretching, and a symmetrical acetylene shows no triple-bond stretch at all. It is, however, seen in the Raman spectrum. Conjugation with carbonyl groups usually has little effect on the position of absorption. When more than one acetylenic linkage is present, and sometimes when there is only one, there are sometimes more absorption bands in this region than there are triple bonds to account for them.

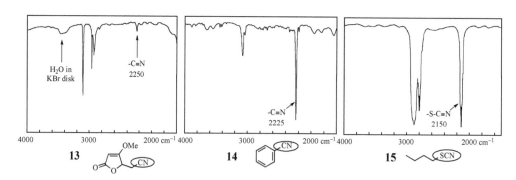

Fig. 2.7 Unusually weak saturated and strong conjugated nitrile absorption

Nitriles absorb in the range 2260-2200 cm^{-1}, but are often weak, as in the spectrum of the nitrile **13**, for example, which was originally assigned the wrong structure because the weak nitrile absorption had been overlooked. Cyanohydrins are notorious in this respect, frequently showing no nitrile absorption at all. As usual, conjugation lowers the frequency and increases the intensity, as in the spectrum of benzonitrile **14**. Isonitriles, nitrile oxides, diazonium salts and thiocyanates such as **15** absorb strongly in the region 2300-2100 cm^{-1} (Table 2.6).

The stretching vibrations of the two double bonds in cumulated double-bonded systems X=Y=Z, such as those found in carbon dioxide, isocyanates, isothiocyanates, diazoalkanes, ketenes and allenes, are strongly coupled, with an unsymmetrical and a symmetrical pair. The former gives rise to a strong band in the range 2350-1930 cm^{-1} and the latter to a band in the fingerprint region, except for symmetrical systems, for which it is IR-forbidden. Carbon dioxide has a sharp band at 2349 cm^{-1}, whereas allenes such as dimethylallene **17** give a sharp absorption in the characteristically narrow range 1950-1930 cm^{-1} (at the lower end of the range). The other cumulated double-bond systems come in between (Table 2.6), with isocyanates and isothiocyanates, such as cyclohexylisothiocyanate **16**, giving rise to a broad band in contrast to the narrow band seen for thiocyanates **15** and allenes **17**.

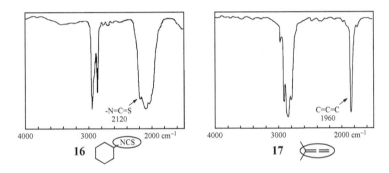

Fig. 2.8 Characteristic absorptions of an isothiocyanate and an allene

2.9 Absorption frequencies of the double-bond region 1900-1500 cm^{-1}

C=O Double bonds (Table 2.7). Identifying which of the several kinds of carbonyl group is present in a molecule is one of the most important uses of infrared spectroscopy. Carbonyl bands are always strong, as in the spectrum of ethyl benzoate **2** in Fig. 2.4, with acids generally stronger than esters, and esters stronger than ketones or aldehydes. Amide absorption is usually similar in intensity to that of ketones but is subject to greater variations.

The precise position of carbonyl absorption is governed by the electronic structure and the extent to which the carbonyl group is involved in hydrogen bonding. The general trends of structural variation on the position of C=O stretching frequencies are summarised in Fig. 2.9.

1. The more electronegative the group X in the system R—C(=O)—X, the higher is the frequency, except that this trend from the inductive effect of X is offset by the effect in the π system of any lone pairs on X. Thus, the inductive effect alone raises the frequency of the absorption—acid fluorides have higher C=O stretching frequencies than chlorides, which are higher than bromides. Similarly, acid chlorides have higher

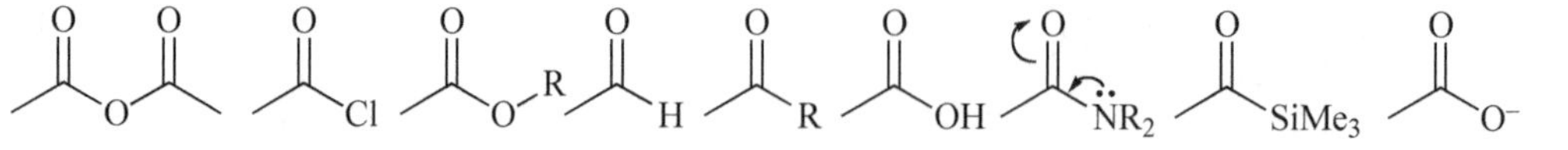

anhydrides	acid chlorides	esters	aldehydes	ketones	acids	amides	acylsilanes	carboxylate ions
1820 & 1760	1800	1740	1730	1710	1710	1660	1640	1580

Fig. 2.9 Representative stretching frequencies (in cm⁻¹) of C=O groups

frequencies than comparable esters, and esters higher than amides. Continuing the series, an electropositive substituent like a silyl group lowers the frequency further still. However, ketones have their C=O stretching frequencies between those of comparable esters and amides, and so the electronegativity of X is not the only factor. The overlap of the lone pair of electrons on X with the C=O bond, illustrated by the curly arrows in the amide in Fig. 2.9, reduces the double-bond character of the C=O bond, while increasing, of course, the double-bond character of the C—N bond. This overlap will have most effect when the lone pair is relatively high in energy. The less electronegative X is, the higher is the energy of its lone-pair orbitals, and the more effective the overlap in reducing the π-bonding in the C=O bond. The net result is to lower the frequency of the C=O stretching vibration in amides below that of ketones, even though the electronegativity of N is higher than that of C. On the other hand, the higher electronegativity of O, coupled with the less-effective π-overlap of its lone pairs, keeps the stretching frequency of esters above that of ketones. Similarly, the hyperconjugative overlap of the neighbouring C—H (or C—C) bonds in a ketone reduces the C=O double-bond character and lowers its stretching frequency relative to that of an aldehyde. At the extreme, carboxylate ions absorb at the lowest frequency of the common carbonyl groups. On the other hand, if the oxygen lone pair in an ester overlaps with another double bond, it is less effective in lowering the frequency of the carbonyl vibration, with the result that vinyl and phenyl esters absorb at higher frequency than alkyl esters, and carboxylic anhydrides even more so. Anhydrides also show two bands rather than one, because the two carbonyl groups have an unsymmetrical and a symmetrical vibration, with the former pushed a little above that of a comparable acid chloride and the latter rather more below it.

2. Hydrogen bonding to a carbonyl group causes a shift to lower frequency of 40-60 cm⁻¹. Acids, amides with an N—H, enolised β-dicarbonyl systems, and o-hydroxy- and o-aminophenyl carbonyl compounds show this effect, as illustrated by the carboxylic acid's coming in the list in Fig. 2.9 between the ester and the amide, instead of being the same as the ester. All carbonyl compounds tend to give slightly lower values for the carbonyl stretching frequency in the solid state compared with the value for dilute solutions.

3. α,β-Unsaturation causes a lowering of frequency of 15-40 cm⁻¹ because overlap similar to that illustrated for the amide in Fig. 2.9, but less effective, reduces the double-bond character of the C=O bond. The unsaturated ketone group in cortisone acetate **1** can be seen as the lowest frequency of the three carbonyl peaks in Fig. 2.2, with the saturated ketone just above it and the ester just above that. The effect of α,β-unsaturation is noticeably less in amides, where little shift is observed, and that sometimes even to higher frequency.

4. Ring strain in cyclic compounds causes a relatively large shift to higher frequency (Fig. 2.10). This phenomenon provides a remarkably reliable test of ring size, distinguishing clearly between four-, five- and larger-membered ring ketones, lactones

and lactams. Six-ring and larger ketones, lactones and lactams, show the normal frequency found for the open-chain compounds.

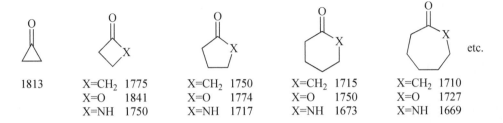

| 1813 | X=CH$_2$ 1775
X=O 1841
X=NH 1750 | X=CH$_2$ 1750
X=O 1774
X=NH 1717 | X=CH$_2$ 1715
X=O 1750
X=NH 1673 | X=CH$_2$ 1710
X=O 1727
X=NH 1669 | etc. |

Fig. 2.10 The effect of ring size on C=O stretching frequencies (in cm^{-1})

5. Where more than one of the structural influences on a particular carbonyl group is operating, the net effect is usually close to additive, with α,β-unsaturation and ring strain having opposite effects.

Amides are special in showing greater variation in the number and position of the bands in the carbonyl region. In particular, they show extra bands in addition to the localised stretching of the C=O bond. Thus, primary and secondary amides, which also have an N—H bond, show at least two bands. The one at higher frequency is more or less the localised stretching vibration of the C=O group; it is called the amide I band. The other, at lower frequency, is more or less the localised bending vibration of the N—H bond; it is called the amide II band. The spectrum of *N*-methylacetamide **18** in Fig. 2.11 shows the amide I and II bands, whereas the spectrum of *N,N*-dimethylacetamide **19** shows a single peak. Both bands are affected by hydrogen bonding and are therefore significantly different when the spectra are taken of solutions or of the solid amide. Lactams, even those with an N—H like caprolactam **12**, do not always show the amide II band.

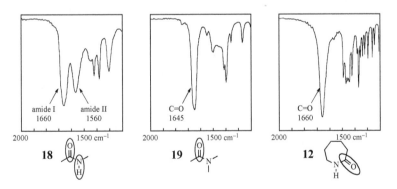

Fig. 2.11 Amide C=O and alkene stretching vibrations

For the full range of all types of carbonyl absorption, see Table 2.7.

C=N Double bonds (Table 2.8). Imines, oximes and other C=N double bonds absorb in the range 1690-1630 cm^{-1}. They are weaker than carbonyl absorption, and are difficult to identify because they absorb in the same region as C=C double bonds.

C=C Double bonds (Table 2.10). Unconjugated alkenes absorb in the range 1680-1620 cm^{-1}. The more substituted C=C double bonds absorb at the high-frequency end of the range, the less substituted at the low-frequency end. The absorption may be very weak or even absent when the double bond is more or less symmetrically substituted, but the vibration frequency can then be detected and measured in the Raman spectrum. Oleyl alcohol **20** has a *cis* double bond near the middle of a long chain. Since it is locally almost symmetrically substituted, its stretching is only just detectable in the infrared spectrum at the top in Fig. 2.12. In contrast, it is seen as a strong peak at 1680 cm^{-1} in the Raman spectrum in grey at the bottom. Raman spectra are usually plotted upwards.

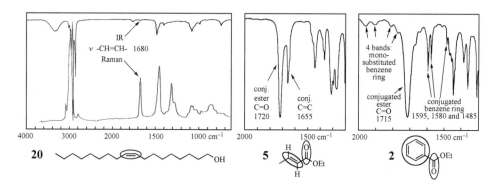

Fig. 2.12 Infrared spectra of alkenes and a benzene ring and a Raman spectrum of an alkene

Note that the infrared spectrum of the alcohol **20** does not have a strong band at 960 cm^{-1}, showing that the double bond is not *trans*. As with carbonyl groups, conjugation lowers the stretching frequency of C=C double bonds, as seen in the spectrum of ethyl crotonate **5** in Fig. 2.12, where it gives rise to the strong peak at 1655 cm^{-1}, which is nevertheless weaker than the carbonyl peak of the conjugated ester at 1720 cm^{-1}. It can also be seen as the small peak at 1615 cm^{-1} on the low-frequency side of the three carbonyl peaks in Fig. 2.2. If the double bond is exocyclic to a ring the frequency rises as the ring size decreases. A double bond within a ring shows the opposite trend: the frequency falls as the ring size decreases. The =C—H stretching frequency rises slightly as ring strain increases, and the =C—H vibration frequencies may give additional structural information.

Aromatic rings. Two or three bands in the 1600-1500 cm^{-1} region are shown by most six-membered aromatic ring systems such as benzenes, polycyclic aromatic rings and pyridines. Typically, a benzene ring conjugated to a double bond has three bands like those marked in the spectrum of ethyl benzoate **2** in Fig. 2.12, usually close to 1600, 1580 and 1500 cm^{-1}. The relative intensities vary, especially as a consequence of the degree of conjugation. Further bands occur in the fingerprint region between 1225 and 950 cm^{-1} which are of little diagnostic value. Aromatic compounds are also characterised by the weak C—H stretching band near 3030 cm^{-1} (Table 2.1), and there are bands in the 1225-950 cm^{-1} region which are of little use in providing structural information. In addition, the shape and number of the two to six overtone and combination bands which appear in the 2000-1600 cm^{-1} region is a function of the substitution pattern of the benzene ring. The pattern of four bands for a monosubstituted ring can be seen in the spectrum of ethyl benzoate **2** in Fig. 2.12. The use of this region for this purpose has been overtaken by

NMR spectroscopy, which does the job much better, but the characteristic patterns for the various substituted benzenes can be found in more specialised books. A fourth group of bands below 900 cm^{-1} produced by the out-of-plane C—H bending vibrations (see Table 2.2) is affected by the number of adjacent hydrogen atoms on the ring. It too can be, but rarely is nowadays, used to identify the substitution pattern.

N=O Double bonds. Nitro groups have unsymmetrical and symmetrical N=O stretching vibrations, the former a strong band in the range 1570-1540 cm^{-1} and the latter a strong band in the range 1390-1340 cm^{-1}, in the fingerprint region. They can be seen easily as by far the strongest peaks in a relatively unfunctionalised compound like *o*-nitrobenzyl alcohol **21**, but they are still evident, emphasised by the black fill-in, in the acetylenic acid **22** in Fig. 2.13, which has a number of other strong peaks, including an unusual carboxylate ion coexisting with the carboxylic acid absorption. The conjugation with the benzene ring has lowered these bands by approximately 30 cm^{-1} into the lower end of the range for nitro groups. Nitrates, nitramines, nitrites and nitroso compounds also absorb in the 1650-1400^{-1} region (Table 2.11).

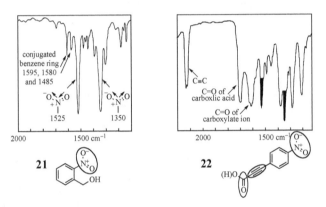

Fig. 2.13 Asymmetrical and symmetrical stretching absorption of nitro groups

2.10 Groups absorbing in the fingerprint region <1500 cm^{-1}

Strong bands in the fingerprint region arise from the stretching vibrations of a few other doubly bonded functional groups like sulfonyl, thiocarbonyl and phosphoryl, but they are easily confused with the stretching vibrations of single bonds like C—O and C—halogen, which are also strong, and the bending vibrations of C—H bonds. They are rarely diagnostically useful, unless their absence is informative. Tables 2.9-2.18 give an indication where some of these absorptions commonly occur, and include some of the less common and more specialised functional groups, like those of boron, silicon, phosphorus and sulfur.

2.11 Internet

The Internet is a continuously evolving system, with links and protocols changing frequently. The following information is inevitably incomplete and may no longer apply, but it gives you a guide to what you can expect. Some websites require particular operating systems and may only work with a limited range of browsers, some require payment, and some require you to register and to download programs before you can use them.

Infrared spectroscopy is better served on the Internet than ultraviolet spectroscopy.

For guides to spectroscopic data on the Internet, see the websites at MIT, the University of Waterloo and the University of Texas, representative of several others. They are tailored for internal use, but are informative nevertheless:

http://libraries.mit.edu/guides/subjects/chemistry/spectra_resources.htm
http://lib.uwaterloo.ca/discipline/chem/spectral_data.html
http://www.lib.utexas.edu/chem/info/spectra.html

The Chemical Database Service is available free to UK academic institutions, and it includes the SpecInfo system from Chemical Concepts, covering IR, NMR and mass spectra. To register go to:

http://cds.dl.ac.uk

Once you have an ID and Password, go to http://cds.dl.ac.uk/specsurf. To learn what is there and how to use the system, click on the online demo link or on the Illustrated Guide link. The service has 21 000 IR spectra. To look for the IR spectrum of a compound you are interested in, go to:

http://cds.dl.ac.uk/specsurf

Click on Start SpecSurf. When the window loads, pull down Edit-Structure to go to a new window in which you draw the structure. Pull down File-Transfer to drop the structure back into the earlier window, and pull down Search-Structure, which will create a list at the bottom of the window of any available spectra for the compound you have drawn. These may be NMR spectra, mass spectra or IR spectra, and unfortunately are not identified as such. The item at the top of the list, usually a ^{1}H-NMR spectrum, will be displayed in the lower spectrum window. Click on each member of the list until the spectrum that appears is an IR spectrum, or until one of them creates a sub-list on the right-hand side, with an itemised list of the kinds of spectra within it. Then click on the IR example, and it will appear in the spectrum window. Hover the cursor over a peak and its frequency in wavenumbers will be displayed in purple just below the spectrum.

The Sadtler database administered by Bio-Rad Laboratories has over 220 000 IR spectra of pure organic and commercial compounds. For information, go to:
http://www.bio-rad.com/ and follow the leads to Sadtler, KnowItAll, and infrared.

Acros provide free access to IR spectra of compounds in their catalogue at:
www.acros.be/

Enter the name or draw the structure of the compound you are interested in, and click on the IR tab.

A website listing databases for IR, NMR and MS is:
http://www.lohninger.com/spectroscopy/dball.html

The Japanese Spectral Database for Organic Compounds (SDBS) has free access to IR, Raman, ^{1}H- and ^{13}C-NMR and MS data at:
http://www.aist.go.jp/RIODB/SDBS/cgi-bin/cre_index.cgi

There is a database of >5000 compounds with gas phase infrared data on the NIST website belonging to the United States Secretary of Commerce:

http://webbook.nist.gov/chemistry/name-ser.html

Type in the name of the compound you want, check the box for IR spectrum, and click on Search, and if the infrared spectrum is available it will show it to you.

Sigma-Aldrich has a library of >60 000 FTIR spectra, access to which requires payment; for information go to:

http://www.sigmaaldrich.com/Area_of_Interest/Equip____Supplies_Home/Spectral_ Viewer/FT_IR_Library.html (that is 4 underline symbols between Equip and Supplies)

ACD (Advanced Chemistry Development) Spectroscopy sell proprietary software called ACD/SpecManager that handles all four spectroscopic methods, as well as other analytical tools:

http://www.acdlabs.com/products/spec_lab/exp_spectra/

It is able to process and store the output of the instruments that take spectra, and can be used to catalogue, share and present your own data. It also gives access to free databases, and to prediction and analysis tools—assigning peaks in the infrared to functional groups, for example:

Wiley-VCH keep an up-to-date website on spectroscopic books and links. The URL for infrared spectroscopy is:

http://www.spectroscopynow.com/Spy/basehtml/SpyH/1,1181,3-0-0-0-0-home-0-0,00.html

and they also provide a link to the SpecInfo databases:

http://www3.interscience.wiley.com/cgi-bin/mrwhome/109609148/HOME

and to ChemGate, which has a collection of 700 000 IR, NMR and mass spectra:

http://chemgate.emolecules.com

There are expensive collections for industrial chemists through IR Industrial Organic Chemicals Vols. 1 and 2, BASF Software, January 2006.

2.12 Bibliography

DATA

The Aldrich Library of FT-IR Spectra, Aldrich Chemical Company, Milwaukee, 1985.

D. Dolphin and A. Wick, *Tabulation of Infrared Spectral Data*, Wiley, New York, 1977.

D. Liu-Vlen, N. B. Colthup, W. G. Fately and J. G. Grasselli, *Handbook of IR and Raman Frequencies of Organic Molecules*, Academic Press, San Diego, 1991.

K. G. R. Pachler, F. Matlock and H.-U. Gremlich, *Merck FT-IR Atlas*, VCH, 1988.

E. Pretsch, P. Bühlmann and C. Affolter, *Structure Determination of Organic Compounds Tables of Spectral Data*, Springer, Berlin, 3rd Ed., 2000.

Sadtler Handbook of Infrared Grating Spectra, Heyden, London.

B. Schrader, *Raman/Infrared Atlas of Organic Compounds*, VCH, 2nd Ed., 1989.

T. J. Bruno and P. D. N. Svoronos, *CRC Handbook of Fundamental Spectroscopic Correlation Charts*, CRC Press, Boca Raton, 2006.

TEXTBOOKS

L. J. Bellamy, *The Infrared Spectra of Complex Molecules*, Chapman & Hall, London, Vol. 1, 1975, Vol. 2, 1980.

K. Nakanishi, *Infrared Absorption Spectroscopy*, Holden-Day, San Francisco, 2nd Ed., 1977.

P. R. Griffiths and J. A. de Haseth, *Fourier Transform Infrared Spectroscopy*, Wiley, New York, 1986.

N. P. G. Roeges, *A Guide to the Complete Interpretation of Infrared Spectra of Organic Structures*, Wiley, Chichester, 1998.

H. Günzler and H.-U. Gremlich, *IR Spectroscopy*, Wiley-VCH, Weinheim, 2002.

P. Hendra, C. Jones and G. Wames, *Fourier Transform Raman Spectroscopy*, Ellis Horwood, Chichester, 1991.

S. F. Johnston, *FT-IR*, Ellis Horwood, London, 1992.

B. C. Smith, *Fundamentals of Fourier Transform Infrared Spectroscopy*, CRC Press, Boca Raton, 1996.

B. C. Smith, *Infrared Spectral Interpretation*, CRC Press, Boca Raton, 1999.

R. L. McCreery, *Raman Spectroscopy for Chemical Analysis*, Wiley-VCH, New York, 2000.

B. H. Stuart, *Infrared Spectroscopy*, Wiley, New York, 2002.

G. Socrates, *Infrared and Raman Characteristic Group Frequencies*, Wiley, Chichester, 2001.

E. Smith and G. Dent, *Modern Raman Spectroscopy: A Practical Approach*, Wiley, New York, 2004.

THEORETICAL TREATMENTS

G. Herzberg, *Infrared and Raman Spectra of Polyatomic Molecules*, Van Nostrand, Princeton, 1945.

G. R. Barrow, *Introduction to Molecular Spectroscopy*, McGraw-Hill, New York, 1964.

B. Schrader, Ed., *Infrared and Raman Spectroscopy*, Wiley-VCH, Weinheim, 1995.

J. M. Chalmers and P. R. Griffiths, Eds., *Handbook of Vibrational Spectroscopy*, in 5 Vols., Wiley, Chichester, 2001, <www.wileyeurope.com/vibspec>

D. A. Long, *The Raman Effect*, Wiley-VCH, New York, 2001.

2.13 Correlation charts

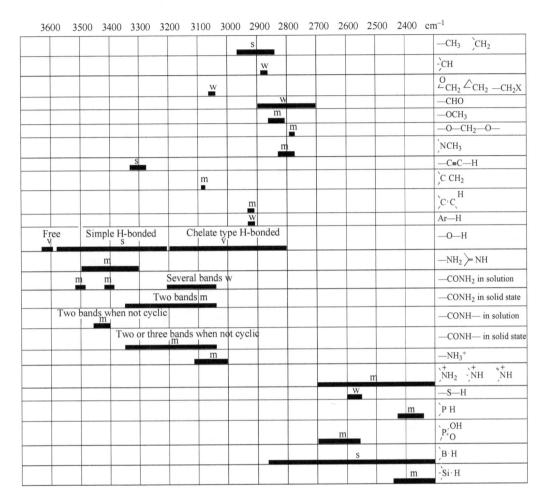

Chart 1 Stretching frequencies of single bonds to hydrogen (Tables 2.1 and 2.3-2.5)

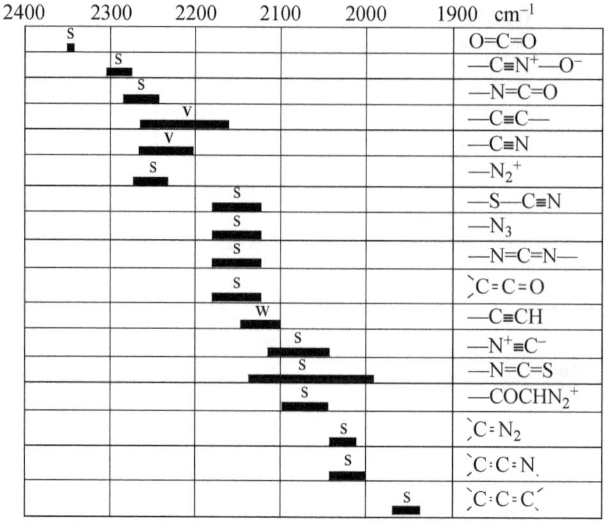

2400	2300	2200	2100	2000	1900 cm⁻¹	
s						O=C=O
	s					—C≡N⁺—O⁻
	s					—N=C=O
	v					—C≡C—
	v					—C≡N
	s					—N₂⁺
		s				—S—C≡N
		s				—N₃
		s				—N=C=N—
		s				C=C=O
		w				—C≡CH
			s			—N⁺≡C⁻
			s			—N=C=S
			s			—COCHN₂⁺
				s		C=N₂
				s		C=C=N
					s	C=C=C

Chart 2 Stretching frequencies of triple and cumulated double bonds (Table 2.6)

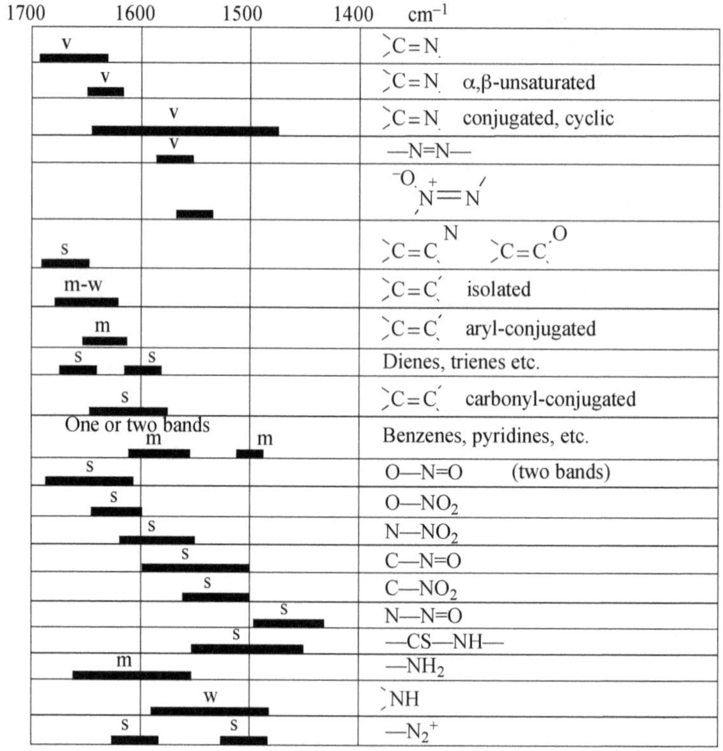

1700	1600	1500	1400 cm⁻¹	
v				C=N
	v			C=N α,β-unsaturated
		v		C=N conjugated, cyclic
	v			—N=N—
		m		⁻O—N⁺=N
s				C=C(N) C=C(O)
m-w				C=C isolated
	m			C=C aryl-conjugated
s		s		Dienes, trienes etc.
	s			C=C carbonyl-conjugated
	m (One or two bands)		m	Benzenes, pyridines, etc.
s				O—N=O (two bands)
	s			O—NO₂
		s		N—NO₂
		s		C—N=O
		s		C—NO₂
			s	N—N=O
			s	—CS—NH—
m				—NH₂
		w		NH
	s		s	—N₂⁺

Chart 3 Stretching frequencies of double bonds other than carbonyl groups (Tables 2.8-2.11) and bending frequencies of N—H bonds (Table 2.4)

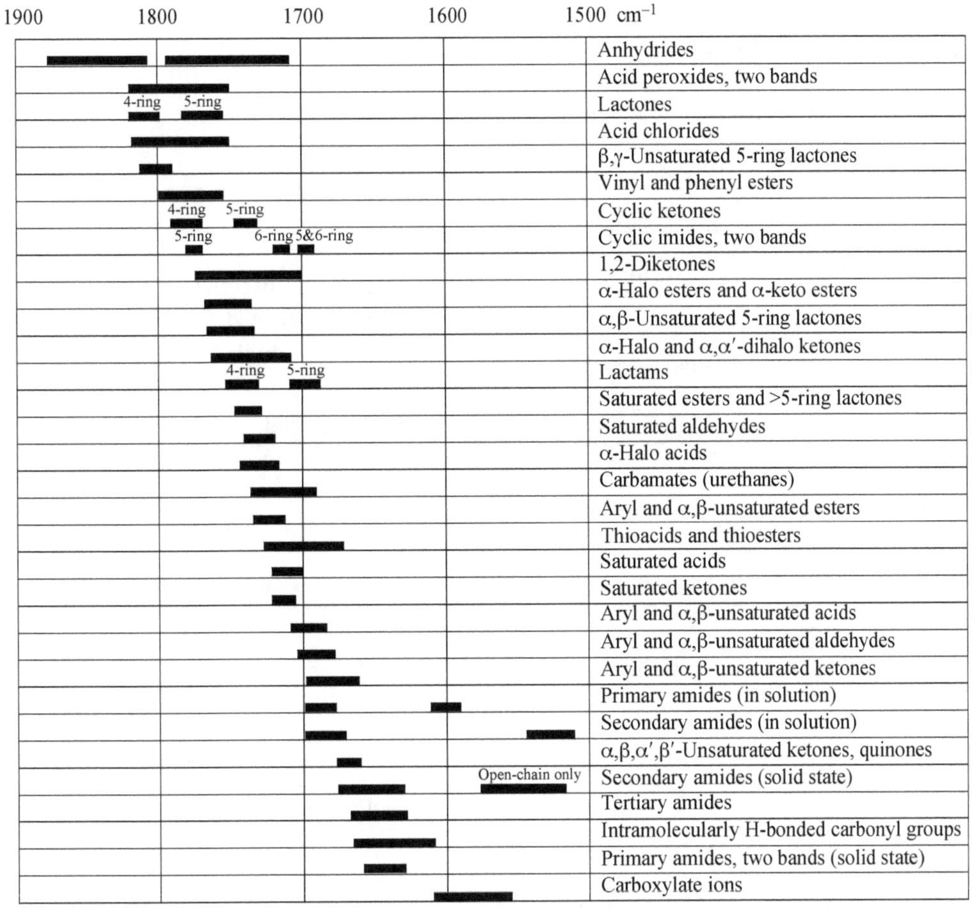
1900
1800
1700
1600
1500 cm^{-1}
Anhydrides
Acid peroxides, two bands
4-ring
5-ring
Lactones
Acid chlorides
β,γ-Unsaturated 5-ring lactones
Vinyl and phenyl esters
4-ring
5-ring
Cyclic ketones
5-ring
6-ring
5&6-ring
Cyclic imides, two bands
1,2-Diketones
α-Halo esters and α-keto esters
α,β-Unsaturated 5-ring lactones
α-Halo and α,α′-dihalo ketones
4-ring
5-ring
Lactams
Saturated esters and >5-ring lactones
Saturated aldehydes
α-Halo acids
Carbamates (urethanes)
Aryl and α,β-unsaturated esters
Thioacids and thioesters
Saturated acids
Saturated ketones
Aryl and α,β-unsaturated acids
Aryl and α,β-unsaturated aldehydes
Aryl and α,β-unsaturated ketones
Primary amides (in solution)
Secondary amides (in solution)
α,β,α′,β′-Unsaturated ketones, quinones
Open-chain only
Secondary amides (solid state)
Tertiary amides
Intramolecularly H-bonded carbonyl groups
Primary amides, two bands (solid state)
Carboxylate ions

Chart 4 Stretching frequencies of carbonyl groups; all bands are strong (Table 2.7)

2.14 Tables of data

Table 2.1 C—H stretching vibrations

Group	Band	Remarks
C≡C−H	~3300(s)	Sharp
$\underset{H}{\overset{H}{C=C}}$	3095-3075(m)	Sometimes obscured by the stronger bands of saturated C—H groups, which occur below 3000 cm^{-1}
$\overset{H}{C=C}$	3040-3010(m)	
Aryl—H	3040-3010(w)	Often obscured
Cyclopropane C—H Epoxide C—H —CH$_2$—halogen	~3050(w)	
—COCH$_3$	3100-2900(w)	Often very weak
Unfunctionalised C—H stretching: CH$_2$ and —CH$_3$	2960-2850(s)	Usually 2 or 3 bands
−CH	2890-2880(w)	
—CHO	2900-2700(w)	Usually 2 bands, one near 2720 cm^{-1}
—OCH$_3$	2850-2810(m)	
NCH$_3$ and NCH$_2$-	2820-2780(m)	
—OCH$_2$O—	2790-2770(m)	

Table 2.2 C—H bending vibrations

Group	Band	Remarks
CH$_2$ and —CH$_3$	1470-1430(m)	Asymmetrical deformations
—C(CH$_3$)$_3$	1395-1385(m) and 1365(s)	
—CH$_3$	1390-1370(m)	Symmetrical deformations
—OCOCH	1385-1365(s)	The high intensity of these bands often dominates this region of the spectrum
C(CH$_3$)$_2$	~1380(m)	A roughly symmetrical doublet
—COCH$_3$	1360-1355(s)	

Table 2.2 continued

Structure	Band	Remarks
C=C (H, H / H, H)	995-985(s) and 940-900(s)	
C=C (H / H)	970-960(s)	C—H out-of-plane deformation. Conjugation shifts the band towards 990 cm^{-1}
C=C (H, H)	895-885(s)	
C=C (H)	840-790(m)	
C=C (H, H)	730-675(m)	C—H out-of-plane deformation
CH_2	~720(w)	Rocking

Table 2.3 O—H stretching and bending vibrations

Group	Band	Remarks
Water in dilute solution	3710	
Water of crystallisation in solids	3600-3100(w)	Usually accompanied by a weak band at 1640-1615 cm^{-1}; residual water in KBr discs shows a broad but weak band at 3450 cm^{-1}
Free O—H	3650-3590(v)	Sharp
H-bonded O—H	3600-3200(s)	Often broad, but may be sharp for some intramolecular single-bridge H-bonds; the lower the frequency the stronger the H-bond
Intramolecularly H-bonded O—H of the chelate type and as found in carboxylic acid dimers	3200-2500(v)	Broad; the lower the frequency the stronger the H-bond; sometimes so broad as to be overlooked
O—H	1410-1260(s)	O—H bending
$-\text{C-OH}$	1150-1040(s)	C—O stretching

Table 2.4 N—H stretching and bending vibrations

Group	Band	Remarks
$-N\overset{H}{\underset{H}{\diagup}}$ $\diagdown N-H$ $=N\overset{H}{\diagup}$	3500-3300(m)	Primary amines show two bands: the unsymmetrical and the symmetrical stretching. Secondary amines absorb weakly, but pyrrole and indole N—H is sharp
Amino acids —NH$_3^+$ Amine salts —NH$_3^+$	3130-3030(m) ~3000(m)	Values for solid state; broad bands also (but not always) near 2500 and 2000 cm^{-1}
$\diagdown\overset{+}{\underset{\diagup}{N}}H_2$ $-\overset{+}{\underset{\diagup}{N}}H$ $\diagdown\overset{+}{N}H$	2700-2250(m)	Values for the solid state; broad because of the presence of overtone bands, etc.
—CONH$_2$	~3500(m) and ~3400(m)	Lowered by ~150 cm^{-1} in the solid state and when involved in H-bonding; often several bands 3200-3050 cm^{-1}
—CONH—	3460-3400(m)	Two bands; lowered in the solid state and when involved in H-bonding; only one band with lactams
	3100-3070(w)	A weak extra band with H-bonded and solid state samples
$-N\overset{H}{\underset{H}{\diagup}}$	1650-1560(m)	N—H bending
$\diagdown N-H$	1580-1490(w)	Often too weak to be noticed
—NH$_3^+$	1600(s) and 1500(s)	Secondary amine salts have the 1600 cm^{-1} band

Table 2.5 Miscellaneous R—H stretching vibrations

Group	Band	Remarks
$\diagdown\underset{\diagup}{P}\overset{\displaystyle O}{\diagdown}_{OH}$	2700-2560(m)	Associated O—H
$\diagdown\underset{\diagup}{B}-H$	2640-2200(s)	
S—H	2600-2550(w)	Weaker than O—H and less affected by H-bonding
$\diagdown P-H$	2440-2350(m)	Sharp

Table 2.5 continued

Group	Band	Remarks
$-\underset{/}{\overset{\backslash}{Si}}-H$ $\underset{H}{\overset{H}{>}}Si$	2360-2150(s) and 890-860 ~2135(s) and 890-860	Sensitive to the electronegativity of substituents
R—D	1/1.37 times the corresponding R—H frequency	Useful when assigning R—H bands; deuteration leads to a recognisable shift to lower frequency

Table 2.6 Stretching frequencies of triple bonds and cumulated double bonds

Group	Band	Remarks
Carbon dioxide O=C=O	2349(s)	Appears in many spectra because of inequalities in path length
Nitrile oxides $-C\equiv\overset{+}{N}-\overset{-}{O}$	2305-2280(m)	
Isocyanates —N=C=O	2275-2250(s)	Very intense; position little affected by conjugation
Internal acetylenes —C≡C—	2260-2150(v)	Strong and at low end of range when conjugated; weak or absent for nearly symmetrical substitution
$R_3SiC\equiv CH$	2040	
Nitriles —C≡N	2260-2200(v)	Strong; conjugated at the low end of the range; occasionally very weak; some cyanohydrins do not absorb in this region
Diazonium salts $-\overset{+}{N}\equiv N$	~2260	
Thiocyanates —S—C≡N	2175-2140(s)	Aryl thiocyanates at the upper end of the range, alkyl at the lower end
Azides $-N=\overset{+}{N}=\overset{-}{N}$	2160-2120(s)	
Carbodiimides —N=C=N—	2155-2130(s)	Very intense; conjugation leads it to split into two bands of different intensity
Ketenes $\underset{/}{\overset{\backslash}{C}}=C=O$	2155-2130(s)	Very intense
Terminal acetylenes —C≡CH	2140-2100(w)	C≡C stretching (ν_{C-H} at ~3300 cm^{-1})

Table 2.6 continued

Group	Band	Remarks
Isothiocyanates —N=C=S	2140-1990(s)	Broad and very intense
Diazoketones —CO-CH=$\overset{+}{N}$=$\overset{-}{N}$	2100-2050(s)	
Diazoalkanes $\diagdown$C=$\overset{+}{N}$=$\overset{-}{N}$	2050-2010(s)	
Ketenimines $\diagdown$C=C=N$\diagdown$	2050-2000(s)	Very intense
Allenes $\diagdown$C=C=C$\diagdown$	1950-1930(s)	Two bands when terminal allene or when bonded to carbonyl and similar groups

Table 2.7 Stretching frequencies of C=O groups (all bands listed are strong)

Group	Band	Remarks
Acid anhydrides —CO—O—CO—		
Saturated	1850-1800 and 1790-1740	Two bands usually separated by about 60 cm^{-1}; the higher-frequency symmetric band is more intense in acyclic anhydrides and the lower frequency asymmetric band is more intense in cyclic anhydrides
Aryl and α,β-unsaturated	1830-1780 and 1770-1710	
Saturated five-ring	1870-1820 and 1800-1750	
All classes	1300-1050	One or two bands; C—O stretching
Acid chlorides —COCl		
Saturated	1815-1790	COF higher, COBr and COI lower
Aryl and α,β-unsat.	1790-1750	
Acid peroxides —CO—O—O—CO—		
Saturated	1820-1810 and 1800-1780	
Aryl and α,β-unsat.	1805-1780 and 1785-1755	

Table 2.7 continued

Group	Band	Remarks
Esters and lactones —CO—O—		
Saturated	1750-1735	
Aryl and α,β-unsaturated	1730-1715	
Aryl and vinyl esters C=C—O—CO—	1800-1750	The C=C stretch is also shifted to higher frequency
Esters with electronegative α-substituents	1770-1745	e.g. (structure: Cl—CH(CH$_3$)—CO—O—)
α-Keto esters	1755-1740	
Six-ring and larger lactones	Similar values to the corresponding open-chain esters	
Five-ring lactones	1780-1760	
α,β-Unsaturated five-ring lactones	1770-1740	When there is an α-C—H present, there are two bands, the relative intensity depending upon solvent
β,γ-Unsaturated five-ring lactones	~1800	
Four-ring lactones	~1820	
β-Keto ester in H-bonding enol form	~1650	Chelate-type H-bond causes shift to lower frequency than normal ester; the C=C is usually near 1630(s) cm^{-1}
Aldehydes	Values below are for solution spectra; lowered by 10-20 cm^{-1} in liquid film or solid state and raised in the gas phase. See also Table 2.1 for C—H	
Saturated	1740-1720	
Aryl	1715-1695	o-Hydroxy or amino groups shift this range to 1655-1625 cm–1 because of intramolecular H-bonding
α-Chloro or bromo	1765-1695	
α,β-Unsaturated	1705-1680	
α,β,γ,δ-Unsaturated	1680-1660	
β-Keto aldehyde in enol form	1670-1645	Lowering caused by intramolecular H-bonding

Table 2.7 continued

Group	Band	Remarks
Ketones	Values below are for solution spectra; lowered by 10-20 cm^{-1} in liquid film or solid state and raised in the gas phase	
Saturated	1725-1705	Branching at the α-position lowers the frequency
Aryl	1700-1680	
α,β-Unsaturated	1685-1665	Often two bands
α,β,α′,β′-Unsaturated and diaryl	1670-1660	
Cyclopropyl	1705-1685	
Six-ring and larger	Similar values to the corresponding open-chain ketones	
Five-ring	1750-1740	α,β-Unsaturation, etc. has a similar effect on these values as on those of open-chain ketones
Four-ring	~1780	
α-Chloro or bromo	1745-1725	Affected by conformation; highest values when halogens are in the same plane as C=O group; α-F has larger effect; α-I has no effect
α,α′-Dichloro or dibromo	1765-1745	
1,2-Diketones s-*trans* (i.e. open-chain α-diketones)	1730-1710	Antisymmetric stretching of both C=O groups; the symmetrical stretch is inactive in IR and active in Raman
1,2-Diketones s-*cis* six-ring	1760 and 1730	In diketo form
1,2-Diketones s-*cis* five-ring	1775 and 1760	In diketo form
1,3-Diketones enol form	1650 and 1615	Lowered by H-bonding and C=C conjugation
o-Hydroxy- or *o*-amino-aryl ketones	1655-1635	Lowering caused by intramolecular H-bonding
Diazoketones	1645-1615	
Quinones	1690-1660	C=C usually near 1600(s) cm^{-1}
Extended quinones	1655-1635	
Tropone	1650	Near 1600 cm^{-1} in tropolones
Carboxylic acids		
All types	3000-2500	O—H stretching; a charateristic group of bands from H-bonding in the dimer

Table 2.7 continued

Group	Band	Remarks
Saturated	1725-1700	Monomer near 1760 cm^{-1}, rarely observed; occasionally both free monomer and H-bonded dimer can be seen in solution spectra; ether solvents give one band ~1730 cm^{-1}
α,β-Unsaturated	1715-1690	
Aryl	1700-1680	
α-Halo	1740-1720	
Carboxylate ions —CO$_2^-$		For special features of amino acids, see text
Most types	1610-1550 and 1420-1300	Asymmetric and symmetric stretching, respectively
Amides and lactams	See Table 2.4 for N—H bands	
Primary —CONH$_2$	~1690 and ~1600	Solution state amide I and II
	~1650 and ~1640	Solid state amide I and II; sometimes overlap; amide I is generally more intense than amide II
Secondary —CONH—	1700-1670 and 1550-1510	Solution state amide I and II; amide II not seen in lactams
	1680-1630 and 1570-1515	Solid state amide I and II; amide II not seen in lactams; amide I is generally more intense than amide II
Tertiary —CONR$_2$	1670-1630	Solid and solution spectra are little different
Five-ring lactams	~1700	Shifted to higher frequency when the N atom is in a bridged system in which overlap of the N lone pair with the C=O π-bond is diminished
Four-ring lactams (β-lactams)	~1745	
—CO—N—C=C		Shifted by +15 cm^{-1} from the corresponding amide or lactam
C=C—CO—N		Shifted by up to +15 cm^{-1} from the corresponding amide or lactam, unusual for the effect of α,β-unsaturation

Table 2.7 continued

Group	Band	Remarks
Imides —CO—N—CO—		
Six-ring	~1710 and ~1700	Shifted +15 cm^{-1} with α,β-unsaturation
Five-ring	~1770 and ~1700	
Ureas N—CO—N		
—NHCONH—	~1660	
Six-ring	~1640	
Five-ring	~1720	
Carbamates (=urethanes)		
—O—CO—N	1740-1690	Also shows an amide II band when non- or mono-substituted on N
—S—CO—N	1700-1670	
Thioesters and acids		
—CO—SH	~1720	Shifted ~25 cm^{-1} when aryl or α,β-unsaturated
—CO—S—alkyl	~1690	Shifted ~25 cm^{-1} when aryl or α,β-unsaturated
—CO—S—aryl	~1710	Shifted ~25 cm^{-1} when aryl or α,β-unsaturated
Carbonates		
—O—CO—Cl	~1780	
—O—CO—O—	~1740	
Ar—O—CO—O—Ar	~1785	
Five-ring	~1820	
—S—CO—S—	~1645	
Ar—S—CO—S—Ar	~1715	
Acylsilanes —CO—SiR$_3$		
Saturated	~1640	
α,β-Unsaturated	~1590	

Table 2.8 C=N; Imines, oximes, etc.

Group	Band	Remarks
$\backslash\!C\!=\!N\!\!\diagdown$ / H	3400-3300(m)	N—H stretching
$\backslash\!C\!=\!N\!\diagdown$	1690-1640(v)	Difficult to identify because of large variations in intensity and the closeness of C=C stretching bands; oximes usually have weak absorptions
α,β-Unsaturated	1600-1630(v)	
Conjugated cyclic systems	1660-1480(v)	

Table 2.9 N=N; Azo compounds

Group	Band	Remarks
—N=N—	1500-1400(w)	Weak or inactive in IR, sometimes seen in Raman
$\backslash\!N\!=\!\overset{+}{N}\!\overset{O^-}{\diagup}$	1480-1450 and 1335-1315	Asymmetric and symmetric stretching

Table 2.10 C=C; Alkenes and arenes

Group	Band	Remarks
$\backslash\!C\!=\!C\!\diagup$	See Table 2.2 for =C—H bands	
Unconjugated	1680-1620(v)	May be very weak if more or less symmetrically substituted
Conjugated with aromatic ring	~1625(v)	More intense than unconjugated C=C
Dienes, trienes etc.	1650(s) and 1600(s)	Lower-frequency band usually more intense and may obscure the higher-frequency band
α,β-Unsaturated carbonyl compounds	1640-1590(s)	Usually weaker than the C=O band
Enol esters, enol ethers and enamines $\overset{O}{\diagup}\!C\!=\!C\!\diagdown$ & $\overset{N}{\diagup}\!C\!=\!C\!\diagdown$	1690-1650(s)	
Aromatic rings	~1600(m)	
	~1580(m)	Stronger when the ring is further conjugated
	~1500(m)	Usually the strongest of the three peaks

Table 2.11 Nitro and nitroso groups, etc.

Group	Band	Remarks
Nitro compounds C—NO_2	1570-1540(s) and 1390-1340(s)	Asymmetric and symmetric N=O stretching; lowered by ~30 cm^{-1} when conjugated
Nitrates O—NO_2	1650-1600(s) and 1270-1250(s)	
Nitramines N—NO_2	1630-1550(s) and 1300-1250(s)	
Nitroso compounds C—N=O		
Saturated	1585-1540(s)	
Aryl	1510-1490(s)	
Nitrites O—N=O	1680-1650(s) 1625-1610(s)	s-*trans* conformation s-*cis* conformation
N-Nitroso compounds N—N=O	1500-1430(s)	
N-Oxides $-\overset{+}{\underset{|}{N}}-O^-$		
Aromatic	1300-1200(s)	Very strong bands
Aliphatic	970-950(s)	
Nitrate ions NO_3^-	1410-1340 and 860-800	

Table 2.12 Ethers

Group	Band	Remarks
$-\overset{|}{\underset{|}{C}}-O-\overset{|}{\underset{|}{C}}-$	1150-1070(s)	C—O stretching
$\overset{\backslash\backslash}{\underset{/}{C}}-O-\overset{/}{\underset{\backslash}{C}}-$	1275-1200(s) and 1075-1020(s)	
C—O—CH_3	2850-2810(m)	C—H stretching
Epoxides $-\overset{O}{\overset{/\backslash}{\underset{|}{C}-\underset{\backslash}{C}}}-$	~1250, ~900 and ~800	

Table 2.13 Boron compounds

Group	Band	Remarks
B—H	2640-2200(s)	
B—O	1380-1310(vs)	
B—N	1550-1330(vs)	
B—C	1240-620(s)	

Table 2.14 Silicon compounds

Group	Band	Remarks
$-\overset{\displaystyle \backslash}{\underset{\displaystyle /}{Si}}-H$	2360-2150(s) and 890-860	Sensitive to the electronegativity of the substituents
$>Si<\overset{\displaystyle H}{\underset{\displaystyle H}{}}$	~2135(s) and 890-860	
SiMe$_n$	1275-1245(s)	Typically sharp at 1260 cm^{-1}
	~840	n = 3
	~855	n = 2
	~765	n = 1
Si—OH	3690(s)	Free O—H
	3400-3200(s)	H-bonded O—H
Si—OR	1110-1000	
R$_3$Si—O—SiR$_3$	1080-1040(s)	
Si—C≡C—	~2040	
Si—CH=CH$_2$	1600 and 1410	
	~1010 and ~960	
Si—Ph	1600-1590	
	1430	Sharp
	1130-1110(s)	Split into two if Ph$_2$
	1030(w) and 1000(w)	
Si—F	1030-820	SiF$_3$ and SiF$_2$ show 2 bands
Si—Cl	625-425	SiCl$_3$ and SiCl$_2$ show 2 bands

Table 2.15 Phosphorus compounds

Group	Band	Remarks
P—H	2440-2350(s)	Sharp
P—Ph	1440(s)	Sharp
P—O—alkyl	1050-1030(s)	
P—O—aryl	1240-1190(s)	
P=O	1300-1250(s)	
P=S	750-580	
P—O—P	970-910	Broad
(P with O and OH)	2700-2560	H-bonded O—H
	1240-1180	P=O stretching
P—F	1110-760	

Table 2.16 Sulfur compounds

Group	Band	Remarks
S—H	2600-2550(w)	Weaker than O—H and less affected by H-bonding
C=S	1200-1050(s)	
(S=C—N)	~3400(m)	N—H stretching, when present; lowered in solid state by H-bonding
	1550-1450(s) and 1300-1100(s)	Amide-II and amide I, respectively
—O—CS—O	~1225	
—O—CS—N	~1170	
N—CS—N	1340-1130	
—S—CS—N	~1050	
—S—CS—S—	~1070	
S=O	1060-1040(s)	
(SO₂)	1350-1310(s)	
	1160-1120(s)	
—SO₂—N	1350-1330(s) and 1180-1160(s)	
—SO₂—O—	1420-1330(s) and 1200-1145(s)	
—SO₂—Cl	1410-1375 and 1205-1170	
—S—F	815-755	

Table 2.17 Halogen compounds

Group	Band	Remarks
C—F	1400-1000(s)	Sharp
	780-680	Weaker bands
C—Cl	800-600(s)	
C—Br	750-500(s)	C—H stretching
C—I	~500	

Table 2.18 Inorganic ions (all bands listed are strong)

Group	Band	Remarks
NH_4^+	3300-3030	
CN^-, ^-SCN, ^-OCN	2200-2000	
CO_3^{2-}	1450-1410	
SO_4^{2-}	1130-1080	
NO_3^-	1380-1350	
NO_2^-	1250-1230	
PO_4^{2-}	1100-1000	

3. Nuclear magnetic resonance spectra

3.1 Nuclear spin and resonance

The phenomenon of nuclear magnetic resonance (NMR) was first observed in 1946, and it has been routinely applied in organic chemistry since about 1960. It has grown steadily in power and versatility, conspicuously since the late 1970s with the introduction of superconducting magnets and Fourier transform (FT) NMR spectroscopy on a routine basis. Although fast becoming a scientific discipline in its own right, it can still be applied to most problems of structure determination without the help of experts.

Some atomic nuclei have a nuclear spin (I), and the presence of a spin makes these nuclei behave like bar magnets. In the presence of an applied magnetic field the nuclear magnets can orient themselves in $2I + 1$ ways. Those nuclei with an odd mass number have nuclear spins of $1/2$, or $3/2$, or $5/2$, ..., etc. (Table 3.4, to be found with other data at the end of this chapter). In the application of NMR spectroscopy in organic chemistry, the ^{1}H and ^{13}C nuclei are the most important, and both have spins of $1/2$. These nuclei, therefore, can take up one of only two orientations, a low-energy orientation aligned with the applied field and a high-energy orientation opposed to the applied field. The difference in energy is given by:

$$\Delta E = \frac{h\gamma B_0}{2\pi} \tag{3.1}$$

where γ is the magnetogyric ratio, which essentially measures the strength of the nuclear magnets and is different for each nucleus, and B_0 is the strength of the applied magnetic field. The number of nuclei in the low-energy state (N_α) and the number in the high-energy state (N_β) will differ by an amount determined by the Boltzmann distribution:

$$\frac{N_\beta}{N_\alpha} = e^{-\frac{\Delta E}{kT}} \tag{3.2}$$

When a radio-frequency signal is applied to the system, this distribution is changed if the radio frequency matches the frequency at which the nuclear magnets naturally precess in the magnetic field B_0: the net effect is that more of the N_α nuclei are promoted from the low-energy state to the high-energy state than in the reverse direction (N_β to N_α), and N_β increases (Fig. 3.1). The resonance frequency in Hz is given by:

$$v = \frac{\gamma B_0}{2\pi} \tag{3.3}$$

and is therefore dependent upon both the applied field strength and the nature of the nucleus in question. The frequencies at which nuclei come into resonance at a field strength of 9.4 Tesla (94 kilogauss) are given in Table 3.4, but for now it is enough to know that at this field strength ^{1}H comes into resonance at 400 MHz, ^{13}C at 100.6 MHz, ^{19}F at 376.3 MHz, ^{29}Si at 79.4 MHz and ^{31}P at 161.9 MHz. Because of the very widespread use of proton NMR spectroscopy, it is usual to refer generally to an instrument with this field strength as a 400 MHz instrument. Virtually all NMR spectrometers in use today have superconducting magnets cooled with liquid helium working with field strengths of this order of magnitude.

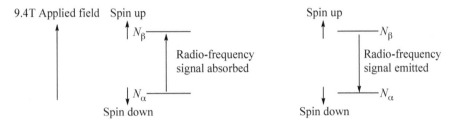

^{1}H 400 MHz, ^{13}C 100 MHz, ^{19}F 376 MHz, ^{29}Si 79 MHz, ^{31}P 162 MHz

Fig. 3.1

The difference between N_α and N_β, determined by the Boltzmann distribution (Eq. 3.2), is very small. The higher the field strength, B_0, the greater the difference between N_α and N_β (Eqs. 3.1 and 3.2), which means that higher-field instruments are inherently more sensitive than the older instruments. For protons in a 60 MHz instrument at 300K, the difference between N_α and N_β is only 1 in 10^6, but in a 400 MHz instrument, it is close to 6 in 10^5. This is still a very small difference, making NMR spectroscopy a relatively insensitive technique compared with UV, IR and mass spectroscopy.

Until the early 1970s, spectra were measured at 100 MHz or less using a permanent magnet, and a continuous radio wave (CW). The magnetic field was varied to scan the range of frequencies to be detected, plotting the spectrum directly as it was being taken, the whole process taking a few minutes. This method has now completely gone out of use, in favour of a Fourier transform (FT) method. A radio frequency signal is applied as a single powerful pulse effectively covering the whole frequency range for the nucleus being examined. The pulse, which lasts for a time (t_p) typically of a few microseconds, generates an oscillating magnetic field (B_1) along the x axis, at right angles to the applied magnetic field (B_0) along the z axis (Fig. 3.2, left). Because of the small difference between N_α and N_β, the sample being investigated has a net magnetisation (M), which is initially aligned in the direction of the applied field. The effect of the pulse is to tip the magnetisation through an angle given by:

$$\Theta = \gamma B_1 t_p \qquad (3.4)$$

Commonly, the time (t_p) is chosen so that Θ is 90°, and such pulses are called $\pi/2$ pulses. The magnetisation, disturbed from its orientation along the z axis, precesses in the xy plane (Fig. 3.2, right), just as a gyroscope precesses when it is tipped out of the axis of the gravitational field. A receiver coil is placed to detect magnetisation oriented along the

y axis. Thus, after the pulse has been applied, the detected signal starts along $+y$ (positive signal), precesses to the x axis (zero signal), then to $-y$ (negative signal), and so on. The decaying signal is detected typically for times of the order of a second. The frequency of oscillation detected is the difference between the NMR resonance frequency and the excitation frequency. Although the excitation frequency is formally a little different from the NMR resonance frequency, the former can still excite the latter since the excitation frequency is applied as such a short pulse (μ_s) that it actually behaves as a spread of frequencies as a consequence of the Heisenberg uncertainty principle.

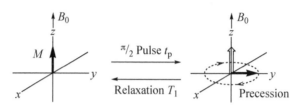

Fig. 3.2

Let us suppose that we are recording the proton NMR spectrum of a simple molecule that contains protons in only two different electronic environments; and that the resonance frequencies of these two protons differ in frequency from the excitation frequency by 5 and 7.5 Hz. The magnetisation in the xy plane arising from the resonance frequency which differs by 5 Hz from the excitation frequency oscillates from positive to negative (and back to positive) 5 times per second. As it does so, it decays exponentially as relaxation gradually allows the nuclear magnets to return to their equilibrium direction along the z axis. The signal is therefore an exponentially decaying cosine of frequency 5 Hz (Fig. 3.3a). Similarly, the signal arising from the resonance which differs from the excitation frequency by 7.5 Hz gives a decaying signal which oscillates 7.5 times per second. In the general case, the signals from the two protons that are in different electronic environments will decay at different rates, and in this example the latter signal has been shown arbitrarily as relaxing more slowly than the former (Fig. 3.3b). By the mathematical manipulation of Fourier transformation (FT), these decaying cosine waves (said to be in the time domain, Fig. 3.3, left) can be converted into frequency signals (said to be in the frequency domain, Fig. 3.3, right). In doing this, the excitation frequency has been taken as the reference frequency, 0 Hz. If the two NMR frequencies are simultaneously excited, then the two decaying cosine waves combine (Fig. 3.3c), but the information necessary to extract both line frequencies is still present. Thus, the decaying signal contains information about a signal which is oscillating relatively quickly and relaxing slowly, and one which is oscillating more slowly and relaxing more quickly (Fig. 3.3, bottom).

   The decaying signals are called *free induction decays* (FIDs). After Fourier transformation of the FID, the spectrum (an FT spectrum) is then plotted, with absorption (upwards) against frequency (increasing towards the left, as in UV and IR spectroscopy). The rate at which the signal decays makes no difference to the appearance of the two lines, provided that the signal has fully decayed by the end of the collection period. The rate of decay can be extracted, and has its uses, but it is not needed in most routine applications of NMR spectroscopy in structure determination. The decaying signals in real NMR spectra are of course much more complicated—spectra typically contain many NMR resonances with a larger range of frequencies.

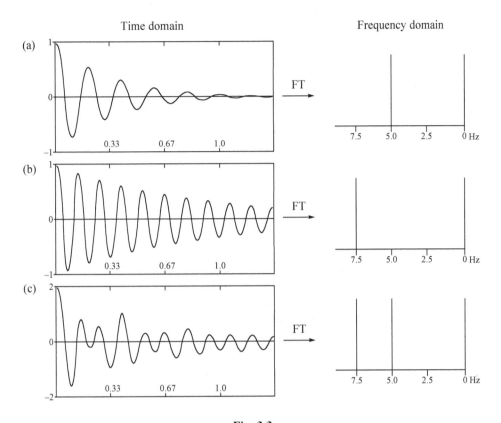

Fig. 3.3

A large number of successive identical pulses can be applied in FT NMR, each followed by an acquisition period, and the FIDs in digital form added together in a computer. Fourier transformation on the sum intensifies the absorption and reduces noise, which is random and largely cancelled out. The accumulation of n spectra (overnight, if necessary for a small sample) improves the signal-to-noise (S/N) ratio by $\sqrt{n}$ relative to that obtained in a single spectrum. With nuclei such as ^{13}C, where this isotope is present in only one in a hundred carbon atoms, and where the nucleus is inherently much less sensitive anyway, there is an even greater need for long runs and for high-field instruments using FT. But, the $\sqrt{n}$ relationship has an important implication for the efficient use of NMR: if the spectrum of a very small and important sample is accumulated for a day, the continuation of the accumulation for a further two days improves S/N only by $\sqrt{3}$. This marginal improvement in S/N for one sample is achieved at the cost of tying up the instrument for a period in which it could have recorded hundreds of spectra of other samples.

In organic chemistry, the most commonly encountered nuclei with $I = \pm^1/_2$ are ^{1}H, ^{13}C, ^{15}N, ^{19}F, ^{29}Si, and ^{31}P. The common nuclei with $I = 0$, ^{12}C and ^{16}O, are inactive. A few other common nuclei have spins, of which ^{2}H and ^{14}N ($I = 1$) are perhaps the most important. When present in organic molecules, they affect ^{1}H and ^{13}C NMR spectra, but it is comparatively unusual to study the NMR spectra of these nuclei themselves in the ordinary course of a structure determination.

3.2 The measurement of spectra

The sample size for a routine ^{13}C spectrum is currently about 30 mg, which will typically require 256 scans over 15 minutes. For a ^1H spectrum 10 mg is enough, and it will typically require 16 scans over one minute. However, with a larger number of pulses, it is possible to obtain high-quality ^{13}C spectra from 1 mg and ^1H spectra from less than 0.1 mg, if the molecular weight is not more than a few hundred. The sample is dissolved in a solvent, preferably one that does not itself give rise to signals in the NMR spectrum. The most commonly used solvent is $CDCl_3$, but polar solvents like d_6-DMSO [$(CD_3)_2SO$] and D_2O are used for polar compounds. The solvent itself provides a signal that is used by the computer to calibrate the spectrum, replacing the older system where an internal standard (usually tetramethylsilane) was added for this purpose. Other solvents, usually per-deuterated, can be used, the choice of solvent being largely determined by the solubility of the compound under investigation. The solution is introduced into a precision-ground pyrex glass or quartz tube to a depth of 2-3 cm. The solution should be free both of paramagnetic and insoluble impurities, and it should not be viscous, or resolution may suffer. The tube is lowered between the poles of the magnet into a probe, which has the transmitter and receiver coils built into it. The magnet is automatically tuned to give the highest possible level of homogeneity, generally about 1 in 10^9, and the tube is spun (at about 30 r.p.s.) about its vertical axis to improve the effective homogeneity even further. With the instrument controlled from a computer keyboard, the pulses are transmitted, the FIDs are accumulated, and stored. The FID is then processed using a computer, and the FT spectrum plotted in the conventional way. The spectroscopic data are also available in digital form as a table of absorption peaks listed by frequency and intensity.

3.3 The chemical shift

The range of frequencies looked at in any one spectrum is a relatively narrow band around the fundamental frequency for the specified nucleus at the field strength of the instrument (Table 3.4). Thus, in ^{13}C spectra—taken, for example, on a 400 MHz instrument—the range of frequencies is a narrow segment of about 20 000 Hz in the neighbourhood of the resonance frequency 100.56 MHz. Within this range each of the different ^{13}C atoms in organic compounds come successively into resonance. The precise frequency at which each carbon comes into resonance is determined not only by the applied field, B_0, but also by minute differences in the magnetic environment experienced by each nucleus. These minute differences are caused largely by the variation in electron population in the neighbourhood of each nucleus, with the result that each chemically distinct carbon atom in a structure, when it happens to be a ^{13}C, will come into resonance at a slightly different frequency from all the others. The electrons affect the microenvironment, because their movement creates a magnetic field. Similarly, in ^1H spectra taken on a 400 MHz instrument, a narrower range—about 4000 Hz in the neighbourhood of the resonance frequency of 400 MHz—is enough to bring all the protons into resonance.

In practice it is inconvenient to characterise the ^{13}C and ^1H peaks by assigning to them their absolute frequency, all very close to 100.56 or 400 MHz: the numbers are cumbersome and difficult to measure accurately. Furthermore, they change from instrument to instrument, and even from day to day, as the applied field changes. It is convenient instead to measure the difference of the frequency (v_s) of the peak from some internal standard (both measured in Hz) and to divide this by the operating frequency in

MHz to obtain a field-independent number in a convenient range. Because the spread of frequencies is caused by the different chemical (and hence magnetic) environments, the signals are described as having a *chemical shift* from some standard frequency. The internal standard almost always used is tetramethylsilane (TMS), and the chemical shift scale δ is then defined by Eq. 3.5.

$$\delta = \frac{v_s (\text{Hz}) - v_{\text{TMS}} (\text{Hz})}{\text{operating frequency (MHz)}} \tag{3.5}$$

The chemical shift δ, which measures the position of the signal, will now be the same whatever instrument it is measured on, whether operating at 200 MHz, 400 MHz or 600 MHz. It has no units and is expressed as fractions of the applied field in parts per million (p.p.m.). Tetramethylsilane is chosen as the reference point because it is inert, volatile, non-toxic, and cheap, and it has only one signal (^{13}C or ^{1}H), which comes into resonance at one extreme of the frequencies found for most carbon and hydrogen atoms in organic structures. By definition it has a δ value of 0 in ^{13}C and ^{1}H spectra. The scale on Fig. 3.4 shows the common range of δ values, which are always written from right to left, a convention that denotes frequencies as higher and positive on the left and lower on the right. Unfortunately, the left and right sides are hardly ever referred to as being at the high- and low-frequency end of the spectrum. Instead, for historical reasons, they are almost invariably referred to by the value of the applied field which corresponds to this relative frequency. The high-frequency end of the spectrum, on the left with high δ values, is described as being *downfield*, and the right side, with low δ values, is said to be *upfield*. It is essential that you get used to this convention (Fig. 3.4).

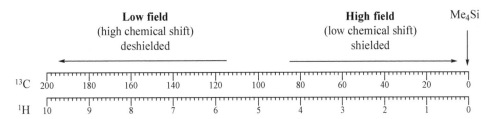

Fig. 3.4 The δ scales for ^{13}C and ^{1}H NMR spectra

Most instruments routinely resolve signals only 0.5 Hz apart. ^{13}C Signals, spread out over 20 000 Hz (in a 100 MHz instrument), rarely coincide, but ^{1}H signals, spread over a narrower range of 4000 Hz (in a 400 MHz instrument) do quite often give coincident, or near coincident signals from different protons. Resolution gets better with instruments working at higher field, but coincidence, or at any rate overlap, is still common.

Figure 3.5 shows the ^{1}H and ^{13}C NMR spectra of 3,5-dimethylbenzyl methoxyacetate **1**, a compound which gives rise to signals over quite a large portion of the usual chemical shift range. In the ^{1}H NMR spectrum, each upward line corresponds to one of the five significantly different kinds of hydrogen atoms, but the signal at δ 6.97 from the hydrogen atom on C-4 has coincided with the signal from the two identical hydrogen atoms on C-2 and C-6. Formally, H-4 is in a different chemical environment from H-2 and H-6, but the environments are so similar that either their chemical shifts are accidentally the same, or are so near to equivalence that they are not resolved. The ^{13}C

NMR spectrum, however, is fully resolved, each upward line corresponding to one of the nine different kinds of carbon atoms. Notice that C-2 and C-6 are in identical environments, because of free rotation about the single bond that joins the aromatic ring to the remainder of the structure. Similarly, C-3 and C-5 are in identical environments, and so are the methyl groups on C-3 and C-5. The solvent in this case is deuterochloroform ($CDCl_3$). In the ^{1}H NMR spectrum, there is a small residual signal at δ 7.25 from the presence of incompletely deuterated chloroform. In the ^{13}C NMR spectrum, on the other hand, the deuterated chloroform is present in large molar excess, it has just as high a proportion of ^{13}C atoms as the ester **1**, and so the solvent signal is relatively intense.

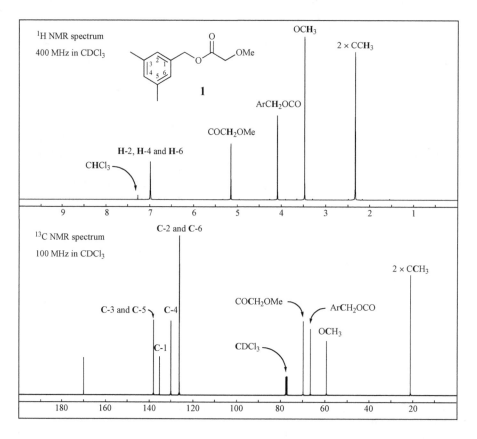

Fig. 3.5

The absorption of a signal in a ^{1}H NMR spectrum is generally proportional to the number of protons coming into resonance at the frequency of that signal, with the result that the area under the peaks is proportional to the number of protons being detected. This is shown in Fig. 3.6, which shows two ways in which the area under the peak can be presented. In the older method, the instrument has plotted an integration trace, starting as a horizontal line and rising from left to right as it passes each absorption. The extent of the rise is proportional to the area under the peak. Measured with a ruler on the trace, and normalised to add up to 16, the numbers, reading from left to right, are 2.91, 2.01, 2.01, 3.00 and 6.07, which shows that the number of hydrogens giving rise to each peak must be, since they can only be integrals, 3, 2, 2, 3 and 6, respectively. More usually, the integration is presented as a number under the peak, with vertical lines indicating the limits of the integral. The actual numbers produced directly by the instrument, based on the assumption that there are three protons in the methoxy signal, are 2.98, 2.05, 2.03, 3.00 and 6.19, which gives some idea of how accurate a routine integration of a reasonably pure sample can be.

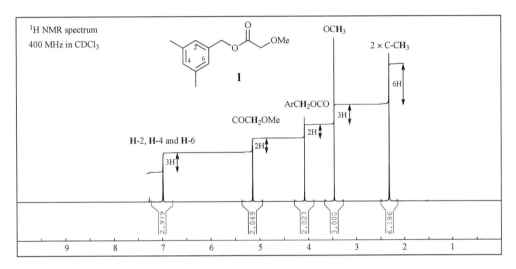

Fig. 3.6

However, for integration to be reliable in FT spectra, all the nuclei must have relaxed to their equilibrium distribution between successive pulses. This is normally the case with protons in ^{1}H spectra, but it is not the case with ^{13}C atoms, which relax much more slowly. Figure 3.5 shows that the peaks in the ^{13}C spectrum are not proportional in intensity to the number of carbon atoms contributing to each signal. For relaxation to occur, the precessing magnetisation of Fig. 3.2 must interact with local fluctuations in the magnetic fields in the molecule, especially those caused by other nuclear magnets. For this reason, the relaxation rate of carbon atoms directly bonded to hydrogen atoms is higher than for carbon atoms not so bonded. This can be seen in Fig. 3.5, where the two lowest-intensity peaks, at δ 170.1 and 135.3, correspond to the carbonyl carbon and C-1, neither of which has a hydrogen atom bonded to it to speed up the relaxation. The other fully substituted carbon atoms, the pair C-3 and C-5, give rise to a signal at δ 138.1 that is approximately twice as strong as the signal from C-1. It is not unusual to see that a pair of identical carbons, C-3 and C-5, gives rise to a signal twice as intense as an otherwise similar carbon—two carbons in a similar situation, will have similar relaxation rates, and hence will give signals more or less in proportion to their abundance. But the signal from C-3 and C-5 at δ 138.1 is not twice as strong as the signal from C-4 at δ 130.1. On the other hand, the signal from C-2 and C-6 at δ 126.3 is about twice as intense as the signal from C-4 at δ 130.1. Because of this variability, integration is not used in routine ^{13}C NMR spectra, except in this limited kind of way, and it is not uncommon to find that some peak intensities are so low (carbonyl groups are notorious in this respect) that they do not appear in the spectrum. The intensity of the ^{13}C signal from deuterochloroform, seen in Fig. 3.5 is not nearly as intense as one might expect, given that the solvent is present in such large excess. The reason for this is that deuterium is much less effective than hydrogen in relaxing the ^{13}C signal. It is possible to increase the intensity of weak or absent signals by adding a paramagnetic salt, which brings with it a powerful magnetic field to speed up the relaxation, but the integration problem is better solved using a multi-pulse technique called gated decoupling.

3.4 Factors affecting the chemical shift

3.4.1 Intramolecular factors affecting the chemical shift

The inductive effect. In a uniform magnetic field, the electrons surrounding a nucleus circulate, setting up a secondary magnetic field opposed to the applied field at the nucleus (Fig. 3.7, where the solid curves indicate the lines of force associated with the induced field). As a result, nuclei in a region of high electron population experience a field proportionately weaker than those in a region of low electron population, and a higher field has to be applied to bring them into resonance. Such nuclei are said to be *shielded* by the electrons.

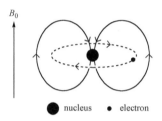

nucleus • electron

Fig. 3.7

Thus, a high electron population shields a nucleus and causes resonance to occur at relatively high field (i.e. with low values of δ). Likewise, a low electron population causes resonance to occur at relatively low field (i.e. with high values of δ), and the nucleus is said to be *deshielded*. The extent of the effect can be seen in the positions of resonance of the ^{1}H and ^{13}C nuclei of the methyl group attached to the various atoms listed in Table 3.1. The electropositive elements (Li, Si) shift the signals upfield, and the electronegative elements (N, O, F) shift the signals downfield, because they donate and withdraw electrons, respectively.

Table 3.1 Chemical shifts for methyl groups attached to various atoms in CH_3X

CH_3X	δ_C	δ_H	CH_3X	δ_C	δ_H
CH_3Li	−14.0	−1.94	CH_3OH	50.2	3.39
$(CH_3)_4Si$	0.0	0.0	$(CH_3)_2S$	19.3	2.09
CH_4	−2.3	0.23	$(CH_3)_2Se$	6.0	2.00
CH_3Me	8.4	0.86	CH_3F	75.2	4.27
CH_3Et	15.4	0.91	CH_3Cl	24.9	3.06
CH_3NH_2	26.9	2.47	CH_3Br	10.0	2.69
$(CH_3)_3P$	16.2	1.43	CH_3I	−20.5	2.13

Hydrogen is more electropositive than carbon, with the result that every replacement of hydrogen by an alkyl group causes a downfield shift in the resonance of that carbon atom and any remaining hydrogens attached to it. Thus, methyl, methylene, methine, and quaternary carbons (and their attached protons) come into resonance at successively lower fields (2-6).

δ_C −2.3	δ_C 8.4	δ_C 15.9	δ_C 25.0	δ_C 27.7
CH_4	$MeCH_3$	Me_2CH_2	Me_3CH	Me_4C
δ_H 0.23	δ_H 0.86	δ_H 1.33	δ_H 1.68	
2	**3**	**4**	**5**	**6**

More dramatically, every replacement of hydrogen by an electronegative element causes a larger downfield shift, more or less additive, in the resonance of the carbon atom and any remaining hydrogens attached to it. Thus, the carbon and the attached protons of methyl chloride, methylene dichloride, chloroform, and carbon tetrachloride come into resonance at successively lower fields (7-10).

δ_C −2.3	δ_C 24.9	δ_C 54.0	δ_C 77.2	δ_C 96.1
CH_4	CH_3Cl	CH_2Cl_2	$CHCl_3$	CCl_4
δ_H 0.23	δ_H 3.06	δ_H 5.33	δ_H 7.24	
2	**7**	**8**	**9**	**10**

Anisotropy of chemical bonds. Chemical bonds are regions of high electron population that can set up magnetic fields. These fields are stronger in one direction than another (they are anisotropic), and the effect of the field on the chemical shift of nearby nuclei is dependent upon the orientation of the nucleus in question with respect to the bond. The anisotropy in π-bonds is especially effective in influencing the chemical shift of nearby protons, as illustrated in Fig. 3.8. When a double bond (Fig. 3.8a) is oriented at right angles to the applied field B_0, the electrons in it are induced to circulate in the plane of the double bond (dashed ellipse in Fig. 3.8a), creating a magnetic field opposed to the applied field at the centre and augmenting it at the periphery. The effect of the induced field is to shift signals from any hydrogens in the augmented region downfield, identified with minus signs (−) in Fig. 3.8 to symbolise deshielding.

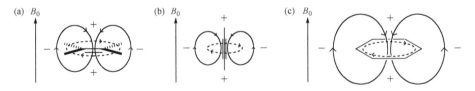

Fig. 3.8

Thus, olefinic hydrogen atoms are shifted downfield relative to their saturated counterparts. But the observed downfield shift is large (Table 3.2), and only partly because of the anisotropy effect; it is also caused by the fact that trigonal carbons are more electronegative (because of their higher s-character) than are tetrahedral carbon atoms. The anisotropic effect also shifts the signals from allylic hydrogens downfield, but to a lesser extent. A carbonyl group has a similar large effect on the hydrogen bonded to it (the case of aldehydes) and a small effect upon the adjacent protons (Table 3.2). Olefinic and carbonyl carbons suffer large downfield shifts relative to their saturated equivalents, although simple models that rationalise these shifts are less easily conveyed. Nevertheless, it is clear that carbonyl carbons and their directly attached protons—which suffer not only the influences experienced by olefinic carbons and their directly attached protons, but also experience the electronegativity of the oxygen atom—exhibit remarkably large chemical shifts (Table 3.2). Triple bonds are noticeably different: they experience an anisotropy set up by electrons circulating around the triple bond with cylindrical symmetry (Fig. 3.8b), and it has the opposite effect from that produced by the π-bond of an alkene or a carbonyl group. The net effect on the chemical shift of acetylenic protons and carbons is for them to come into resonance in between the protons and carbons of alkenes and alkanes (Table 3.2).

Table 3.2 Chemical shifts for carbon and hydrogen in, on and near multiple bonds

Compound	δ_C	δ_H	Compound	δ_C	δ_H
CH_3H	−2.3	0.23	CH_3CHO	31.2	2.20
$CH_3CH=CH_2$	22.4	1.71	CH_3COMe	28.1	2.09
$CH_3C\equiv CH$	3.7	1.80	CH_3CN	1.30	1.98
$CH_2=CH_2$	123.3	5.25	$MeCHO$	199.7	9.80
$HC\equiv CMe$	66.9	1.80	Me_2CO	206.0	-
$MeC\equiv CMe$	79.2	1.75	$MeCN$	117.7	-

An even larger anisotropic effect is produced by the π-system of aromatic rings. The circulating electrons are now called a *ring current*, and they create a relatively strong magnetic field (Fig. 3.8c). The effect of the induced field is to deshield substantially the hydrogens attached to the aromatic ring **12**, which generally come into resonance 1.5-2 p.p.m. downfield from the corresponding olefinic signals in ethylene **11**. The ^{13}C signal for benzene is similarly shifted downfield (by 5.2 p.p.m.), but the effect is much less noticeable because of the relatively large chemical shifts in ^{13}C spectra. In the special case of macrocyclic aromatic rings like [18]-annulene **13**, there are 'inside' as well as 'outside' hydrogens. The 'inside' hydrogens experience a weaker field than the applied field (the + signs in Fig. 3.8c signifying shielding) and come into resonance at a conspicuously high field, while the 'outside' hydrogens experience a stronger field and come into resonance at the low-field end of the aromatic region. Cyclic conjugated systems with 4n electrons are much less common, because they are usually unstable.

11

12

13

They are called antiaromatic, and, when they can be isolated, their ring current circulates in the opposite direction, leading to upfield shifts in the outside protons and downfield shifts in inside protons. The magnetic field induced in double bonds, and especially in aromatic rings, can have a profound effect on the chemical shift of nuclei held anywhere in their neighbourhood, not just those in and attached directly to them. The precise orientation can make a large difference, so that the NMR spectra of complex structures with aromatic rings in them can be hard to predict.

Cyclopropanes present a special case that may be a consequence of a ring current. The carbon and proton resonances of cyclopropanes appear at exceptionally high field, above the usual range for methylene groups, and even above the range for most methyl groups. In cyclopropane itself **14**, the three *cis* vicinal C—H bonds are able to conjugate with each other, just as p-orbitals conjugate with each other. The cyclic six-electron conjugated system thus set up gives rise to a ring current, and both the carbons and the protons sit in the shielding region of the magnetic field induced by that ring current. Substitution by alkyl groups and by electronegative elements moves the resonances downfield in the usual way. The antiaromatic conjugation of four C—H bonds in cyclobutane **15** is much less effective, because of puckering in the ring, but the effect is in the opposite direction—cyclobutane protons come into resonance at slightly lower field than comparable methylene protons, as in cyclopentane **16**, which has chemical shifts similar to those of open-chain compounds.

14   15   16

Polar effects of conjugation. When a double bond carries a polar group, the electron distribution is displaced. The displacement is usually understood as a combination of inductive effects, which operate in the σ-framework (and simply fall off with distance) and conjugative effects, which operate in the π-system (and alternate along a conjugated chain). The effects in the π-system can be illustrated simplistically with curly arrows on the canonical structures for methyl vinyl ether **17** and methyl vinyl ketone **18**. The curly arrows illustrate the effect on a C=C double bond with a π-donor **17** and a π-acceptor group **18**, but molecular orbital calculations of the electron distribution in π-systems support this simple picture.

17   18

These displacements of electron population naturally affect the position of resonance of nearby nuclei, as shown in Table 3.3. In general, π-donor groups (Me < MeO < Me$_2$N) on π-systems shield the β-nuclei, as implied by the canonical structure on the right for **17**, causing an upfield shift relative to their position in ethylene. Largely because of the inductive effect, electronegative elements simultaneously induce a downfield shift of the α nuclei. The effects of π-acceptor groups (Li, SiMe₃ and COMe) on π-systems are not so easily explained using inductive effects and resonance structures like **18**. Because of the way the substituent interacts with the π-system, affecting the anisotropic field surrounding the π-bond, electropositive substituents give rise to downfield shifts on trigonal carbons, both α and β, in contrast to the upfield shifts they induce on tetrahedral carbons.

Table 3.3 Conjugative effects on the chemical shifts of substituted alkenes

X	*Electronic nature*	$\delta_{C\beta}$	$\delta_{C\alpha}$	$\delta_{H\beta}$	$\delta_{H\alpha}$
H	Reference compound	123.3	123.3	5.28	5.28
Me	Weak π- and σ-donor	115.4	133.9	4.88	5.73
OMe	π-Donor, σ-acceptor	84.4	152.7	3.85	6.38
Cl	σ-Acceptor, weak π-donor	117.2	125.9	5.02	5.94
Li	π-Acceptor, σ-donor	132.5	183.4		
SiMe₃	π-Acceptor, σ-donor	129.6	138.7	5.87	6.12
CH=CH₂	Simple conjugation	130.3	136.9	5.06	6.27
COMe	π-Acceptor, weak σ-acceptor	129.1	138.3	6.40	5.85

When the β carbon is unsubstituted, there are two β-hydrogen atoms, and they can experience different fields (the data in Table 3.3 are for the hydrogen *trans* to the substituent). A donor substituent, as in ethyl vinyl ether **20**, causes the β-hydrogen atom *trans* to the substituent to move further upfield, relative to ethylene **19**, than the hydrogen atom *cis* to it. A carbonyl group, as in methyl vinyl ketone **21**, causes the β-hydrogen *cis* to the carbonyl group to be shifted further downfield than the hydrogen *trans* to it, probably because of a direct contribution through space from the anisotropic field induced in the carbonyl group.

Polar groups attached directly to a benzene ring cause upfield and downfield shifts more or less in the same way as they do on a simple double bond. The effects of a π-donor and a π-acceptor group are seen in the structures **23** and **24**, where the signals of the *ortho* and *para* carbons and hydrogens are shifted upfield by the methoxy group, relative to the signals of benzene **22**. The effect of the nitro group is less straightforward, just as the electron-withdrawing groups are on an alkene: the *ortho* hydrogen and the *para* carbon and hydrogen are shifted downfield, as one might expect, but the *ortho* carbon is shifted upfield.

		δ_C	δ_H			δ_C	δ_H
	i	158.7			i	148.1	
	o	113.8	6.81		o	123.2	8.21
	m	129.4	7.17		m	129.3	7.45
	p	120.4	6.86		p	134.5	7.66

22   **23**   **24**

Very approximately, a change in substitution pattern in any organic structure has a similar effect on both the carbon and the proton spectra: the δ value of the carbon signal is about 20 times the δ value of the proton signal. However, there are many large deviations from this general picture, including the opposite effect of the nitro group on the *ortho* carbons and hydrogens of nitrobenzene **24**, and the big effect of the aromatic ring current on protons compared with the appearance of a small effect on the carbon atoms.

Van der Waals forces. When a substituent is pressed close to a hydrogen atom (closer than the sum of their Van der Waals radii), the electron population around the proton is pushed away. The proton is overall deshielded, and comes into resonance at unusually low field, as in the ^{1}H NMR spectrum of 2-adamantanol **25**, where the signal from the two protons H_b appears downfield from the broad, largely unresolved signal (δ 1.89-1.69) from all the other CHs, except, of course, for H_c, which is bonded to a carbon carrying an electronegative element. The proximity-induced displacement of the electrons has the opposite effect on a second proton if there is one bonded to the same carbon atom. This proton is shielded, and as a result the signal from the two protons H_a in adamantanol appears upfield of all the other CHs.

25

The overall pattern that has emerged from the discussion so far is summarised in Fig. 3.9, which gives the approximate ^{1}H and ^{13}C chemical shifts in and adjacent to many of the common functional groups.

Isotope effect. Replacing a lighter isotope by a heavier one causes an upfield shift in the signals from nearby atoms. The effect is only important when a proton is replaced by a deuterium atom, and is only substantial for the ^{13}C chemical shift of the carbon atom to which the deuterium is attached and of the next carbon in the chain. The effect is an easily resolved upfield shift, typically of 10-30 Hz in the directly attached carbon (0.1-0.3 p.p.m. on a machine operating at 100 MHz). The shift for the next carbon in the chain is about one-third as great. These shifts are especially useful in biosynthetic studies, where the carbon atoms attached to deuteriums are easily picked out by the difference in their chemical shift.

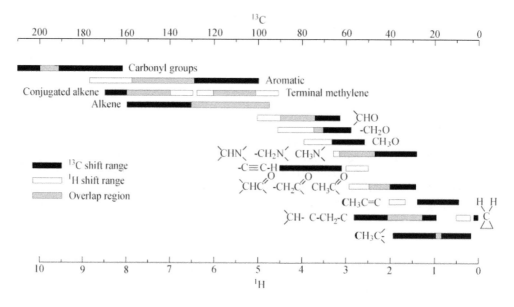

Fig. 3.9

Estimating a chemical shift. The inductive, conjugative and anisotropic effects in polyfunctional molecules are more or less additive, so that we can, for example, see that the proton signals in the ^{1}H spectrum of Fig. 3.5 are in appropriate places. The signal from the C-3 and C-5 methyl groups (at δ 2.31) appears downfield from that of the methyl group in propene (δ 1.71, Table 3.2), because the methyl groups are adjacent to a benzene ring and suffer some of the effect of the ring current, which is typically responsible for a downfield shift of about 0.6 p.p.m. The methylene group attached to C-1 of the benzene ring simultaneously suffers the effects of being benzylic and the effect of being adjacent to an electronegative element. We can expect it to be downfield by about 1.4 p.p.m. for the effect of being benzylic and about 2.8 p.p.m. for being next to an ester oxygen (see Table 3.19 at the end of this chapter for a tabulation of the effects of having various functional groups attached to methyl, methylene and methine carbons), making a total of 4.2 p.p.m. downfield from the position of a simple methylene group (δ 1.33, **4**). Thus, we can expect it to give a signal at δ 5.5, and it actually comes into resonance at δ 5.11. Similarly, using the same Table 3.19 at the end of this chapter, the methylene group between the carbonyl group and the methoxy group might be expected to be shifted downfield by the carbonyl group by 0.9 p.p.m. and by the methoxy group by 2.1 p.p.m., giving an estimate of δ 4.3, close to the observed value of δ 4.05. The methoxy group (at δ 3.45) is a little downfield of the position for methanol (at δ 3.39, Table 3.2), principally because of the anisotropic effect of the nearby carbonyl group. Finally, the aromatic protons at δ 6.97 are a little upfield of the signal from benzene (δ 7.27), because they are *ortho* to two alkyl groups and *para* to another. Alkyl groups, which are mildly electron-donating, shift an *ortho* or *para* proton upfield by about 0.2 p.p.m. (Table 3.25).

But these are very crude estimates, and there are several sets of better empirical rules with which to estimate the chemical shifts commonly encountered for ^{13}C and ^{1}H nuclei. These rules, and the tables of data needed to apply them, are all grouped together at the end of this chapter, where they can easily be found when you use this book as a handbook in the laboratory. Using them, you can estimate the chemical shift of the different kinds of carbon atoms in simple aliphatic compounds (Eq. 3.16 and Tables 3.6, 3.7 and 3.8), in simple alkenes (Eq. 3.17 and Table 3.10), and in polysubstituted benzene rings (Eq. 3.18 and Table 3.12), and of the carbon atoms in the various kinds of carbonyl groups (Table 3.13). There are similar rules and tables for estimating the proton chemical shifts of substituted alkanes (Eq. 3.20 and Table 3.20), substituted alkenes (Eq. 3.21 and Table 3.22), and substituted benzene rings (Eq. 3.22 and Table 3.25), although estimates for protons are rarely as good as those for ^{13}C nuclei.

It is better still, and easier if you have access to suitable computer programs, to draw the chemical structure, and have the program tell you what the probable chemical shifts will be. These programs use similar equations and reference data to those given at the end of this chapter, and save you from doing the sums. The ^1H and ^{13}C chemical shifts for the ester **1**, measured by the spectrometer and estimated by the program ChemNMR incorporated into ChemDraw® Ultra, are shown in Fig. 3.10. This relatively simple compound is unlikely to have unexpected long-range effects, and as a result the estimates are as good as these programs get.

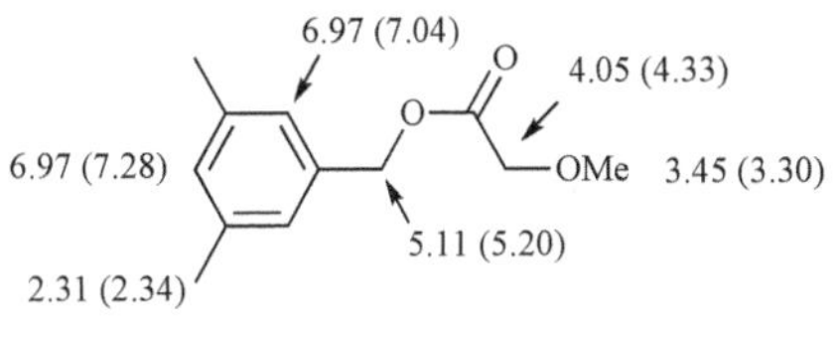

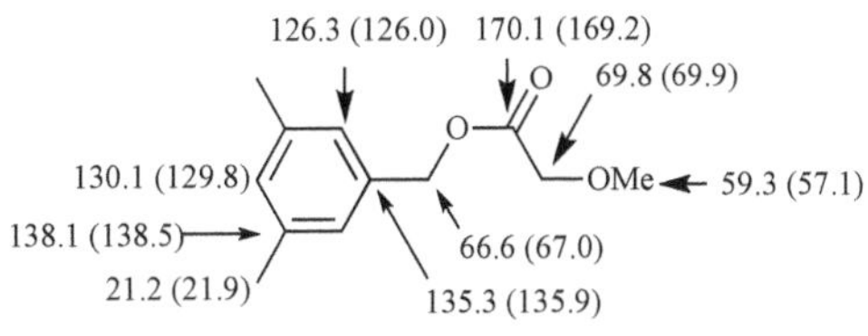

Fig. 3.10

In general, ChemNMR gives ^{13}C chemical shifts with a standard deviation of 2.8 p.p.m. for 95% of compounds, and ^1H chemical shifts, rather less reliably, with a standard deviation of 0.3 p.p.m. for 90% of compounds. The conformation of one molecule may not be the same as the conformation of the model on which the rules are based; the anisotropy of the field then causes the local field in the compound under investigation to differ from that of the model. The effects of distant groups are not included in the rules, usually because they are relatively unimportant, but, in some molecules, a distant group may fold back into the region of the nucleus under investigation and shift its resonance dramatically. This is especially the case when aromatic rings, with their powerful ring currents, are present. Nevertheless, these programs are getting better and better.

3.4.2 Intermolecular factors affecting the chemical shift

Hydrogen bonds. A hydrogen atom involved in hydrogen bonding is sharing its electrons with two electronegative elements. As a result, it is itself deshielded, and comes into resonance at low field. In water, in a very dilute solution in $CDCl_3$, hydrogen bonding is at a minimum for an OH group, and the protons come into resonance at $\delta \sim 1.5$, as can be seen in the ^1H spectrum of Fig. 3.5, which has a minute signal at δ 1.66. This signal is very weak here, because the spectrum was taken with plenty of compound, but it can often be seen more clearly in FT spectra taken with many scans on small samples in more dilute solution. In droplets of water, on the other hand, suspended in $CDCl_3$, the molecules are hydrogen bonded intermolecularly, and they come into resonance at $\delta \sim 4.8$. The position of resonance of the OH and NH protons of alcohols and amines is unpredictable, because the extent to which the hydrogen atoms are involved in hydrogen bonding is both unpredictable and concentration dependent. The usual range is δ 0.5-4.5 for alcohols, δ 1.0-4.0 for thiols, and δ 1.0-5.0 for amines (Table 3.27). The much stronger intermolecular hydrogen bonding in carboxylic acid dimers **26** leads to very low-field absorption in the δ 9-15 range, and the corresponding intramolecular hydrogen bonding of enolised β-diketones **27** and *o*-hydroxycarbonyl compounds **28** is similar (δ 15.4 for acetylacetone **27** itself). These are off the scale of the usual ^{1}H NMR print-out, and may have to be looked for specially (Table 3.27).

| 26 | 27 | 28 |

Fortunately, it is easy to identify the signal from a hydrogen atom bonded to an electronegative element, in spite of the uncertainty about where it will appear in a ^{1}H NMR spectrum: if the sample in CDCl$_3$ solution is shaken with a drop of D$_2$O the OH, NH, and SH hydrogens exchange rapidly with the deuterons, the HDO floats to the surface, out of the region examined by the spectrometer, and the signal of the OH, NH, or SH disappears from the spectrum (or, quite commonly, is replaced by a weak signal close to δ 4.8 from suspended droplets of HDO). This technique is known as a D$_2$O shake.

Temperature. The resonance position of most signals is little affected by temperature, although OH, NH, and SH protons resonate at a higher field at higher temperatures because the degree of hydrogen bonding is reduced.

Solvents. Chemical shifts are little affected by changing solvent from CCl$_4$ to CDCl$_3$ (±0.1 p.p.m.), but change to more polar solvents—such as acetone, methanol, or DMSO—does have a noticeable effect, ±0.3 p.p.m. for ^{13}C and ±0.3 p.p.m. for protons. Benzene can have an even larger effect, ± 1 p.p.m. for protons and for ^{13}C, because it weakly solvates areas of low electron population; since the benzene has a powerful anisotropic magnetic field (Fig. 3.8c), solute atoms lying to the side of or underneath the solvating benzene ring can experience significant shielding or deshielding relative to their position in an inert solvent like CDCl$_3$. Pyridine is sometimes even more effective. This solvent-induced shift can be useful when you want to resolve two signals which overlap in the first spectrum that you take. More substantial shifts can be induced by complexation with paramagnetic salts, as discussed later in Sec. 3.12.

The common solvents in NMR spectroscopy are used in deuterated form, in order not to introduce extra signals, but most of them have residual signals from incomplete deuteration. It is important to recognise these signals, listed in Table 3.26, in order to discount them from spectra that you are interpreting. The weak signal of CHCl$_3$ (at δ 7.25 in ^{1}H spectra) is evident in many of the spectra taken in CDCl$_3$ used to illustrate this chapter. Fortunately, most of them introduce only one or two sharp signals, and they are easily recognised. The carbon of CDCl$_3$ is equally recognisable: it can also be seen in many of the spectra used to illustrate this chapter as the group of three very weak lines centred at δ 77.3. The reason that this signal has three lines, and not one, takes us into the next section.

3.5 Spin-spin coupling to ^{13}C

We have left out of the discussion so far an important effect that neighbouring magnetic nuclei have on the signal we detect. If a nearby nucleus has a spin, that spin affects the magnetic environment of the nucleus we are observing, and the signal we detect is not a single peak, but a group of peaks, the complexity of which is dependent upon the nature and number of the nearby atoms.

3.5.1 ^{13}C-^{2}H Coupling

The carbon atom of $CDCl_3$ is attached to a deuterium nucleus. Deuterium has a spin $I=1$, which means that there are three possible energy levels for a deuterium atom placed in a magnetic field. The carbon atom therefore experiences three slightly different magnetic fields depending upon the spin state of the deuterium nucleus to which it is attached. Since the difference in energy between the three states is very small, there is an essentially equal probability that a carbon atom will be bonded to a deuterium in any one of the three states. The result is that the carbon nucleus comes into resonance at three frequencies with equal probability, as we can almost see in the ^{13}C spectrum in Fig. 3.5, where the $CDCl_3$ signal is actually three equally spaced lines at δ 77.6, 77.25 and 76.9. The carbon is said to be coupled to the deuterium, and the separation of the lines in Hz is called the coupling constant, J. Because there is only one bond between the carbon and the deuterium, J is further qualified as $^1J_{CD}$. Carbon-deuterium coupling is much less important than carbon-hydrogen coupling, but we begin with it because it is visible in all ^{13}C spectra taken in $CDCl_3$. It is also visible in biosynthetic and other studies taking advantage of the isotope shift obtained by attaching a deuterium atom to a ^{13}C nucleus (Sec. 3.4.1). The signal from the ^{13}C atom attached to a deuterium atom is not only shifted upfield by a small but detectable amount but is also distinctively a triplet.

3.5.2 ^{13}C-^{1}H Coupling

Why is there no comparable carbon-hydrogen coupling in the ^{13}C spectrum in Fig. 3.5? The answer is that while that spectrum was being taken, the sample was irradiated with a strong signal encompassing the whole range of frequencies within which the protons in the molecule came into resonance. This caused the N_α and N_β protons to be exchanging places rapidly several times during the measurement of the carbon signal. Each carbon atom, therefore, 'saw' only an average state for the protons near to it; instead of being coupled, each ^{13}C atom simply gave rise to a single sharp line. ^{13}C Spectra are usually taken in this way, and are described as *proton decoupled*.

If we look at the same spectrum without decoupling the protons, we see the spectrum in Fig. 3.11, which has many more lines. The spectrum is more complicated to analyse, and we see why proton decoupling is standard practice when taking ^{13}C spectra. Nevertheless, in this particular case, we can identify the signals from each carbon, and we can identify the multiplicity. The three carbon atoms to which no hydrogen is attached are still singlets, the two different kinds of carbon atoms to which one hydrogen atom is attached are doublets, the two carbon atoms to which two hydrogens are attached are triplets and the carbon atoms to which three hydrogen atoms are attached are quartets.

The easiest signals to interpret are the carbon atoms to which no hydrogen is attached, one from the carbonyl carbon at δ 170.1, one from the pair of identical carbons, C-3 and C-5, at δ 138.1, and one from C-1 at δ 135.3: they are singlets because they have no

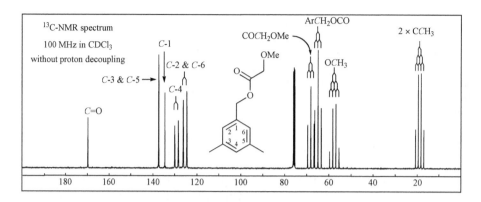

Fig. 3.11

hydrogens on them to couple with. The signals from the other carbon atoms in the aromatic ring, C-4 and the identical pair C-2 and C-6, are at δ 130.1 and 126.3. They have only one hydrogen atom attached to them, and in each case the hydrogen atom ($I = {}^1/_2$) can take up two orientations with respect to the applied field, with essentially equal probability. The carbon atom therefore experiences two slightly different magnetic fields, and comes into resonance as two lines, which is referred to as a doublet (Fig. 3.12b). The two lines are separated by the difference in the resonance frequency, which is measured in Hz and is called the *coupling constant, J*. The two lines of each doublet are equally intense, as you can see in both doublets, but the doublet at δ 126.3 is approximately twice as intense as the doublet at δ 130.1.

The methylene groups in the ester **1** have two protons attached to each carbon. The easiest way to understand the consequence of having two neighbouring protons is to look at the effect of each in turn. The first proton would split the signal into two, and the second would then split it again by the same amount, as illustrated in Fig. 3.12c. You can see, from simple geometry, that the consequence of having two equal coupling constants is a coincident line in the centre. The central line is therefore twice as intense as the two outer lines, and the resulting signal is called a 1:2:1 triplet. In Fig. 3.11 we can see the two triplets centred, like the corresponding lines in the decoupled spectrum in Fig. 3.5, at δ 69.8 and 66.6.

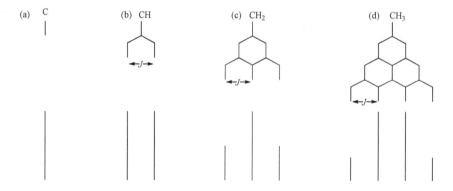

Fig. 3.12

Methyl carbons are each attached to three hydrogens. We can simply extend the argument of Fig. 3.12c to Fig. 3.12d: the first hydrogen splits the carbon into a doublet, the second splits each line of the doublet into a doublet, making a triplet as before, and the third splits each of the lines of the triplet into doublets, creating a quartet. Because the three hydrogens are identical, the coupling constants are identical, and the two central lines are therefore made up of perfect coincidences. The distribution of intensity within the signal creates a 1:3:3:1 quartet, which can be seen in the OCH_3 and the pair of identical CCH_3 signals in Fig. 3.11. In each of the signals shown in Fig. 3.11, the true chemical shift for that carbon atom is the centre of the multiplet.

In summary, the signals of quaternary, methine, methylene, and methyl carbons are a singlet, a doublet, a triplet and a quartet, respectively, Fig. 3.12a-d. These patterns are quite general, and we shall meet them again in proton NMR spectra. The rule, for nuclei of $I = 1/2$, is that a nucleus, equally coupled to n others, gives rise to a signal with $(n + 1)$ lines, and the intensities are given by the coefficients of the terms in the expansion of $(x + 1)^n$ (Table 3.31).

Clearly, the information contained in these multiplets is valuable, but interpretation can be complicated by overlapping signals in complicated molecules, where it becomes nearly impossible to disentangle the multiplets. Even in the fully resolved spectrum of Fig. 3.11, the two triplets almost overlap. The situation can be saved by another technique: while the ^{13}C spectrum is being measured, the sample is irradiated at a frequency close to but not coinciding with the resonance frequency of the protons. This is called *off-resonance decoupling*, and has the effect of narrowing the multiplets, without removing them altogether, as in fully decoupled spectra. Thus, it is easily possible to find out how many of each kind of carbon is present in a molecule of unknown structure. This technique, however, may still produce overlapping multiplets if the molecule is large or has several similar carbon atoms. It has now been superseded by other techniques (see Sec. 3.15 later in this chapter) which completely overcome this problem, and you are not likely to come across off-resonance decoupling except in spectra taken in the past, where you will see the signals reported as singlets, doublets, triplets and quartets.

The size of the coupling constant, J, is also informative. It is principally affected by the geometry of the bonds around the carbon atom, tetrahedral (sp^3) carbon usually giving values between 120 and 150 Hz, trigonal (sp^2) carbon between 155 and 205 Hz, and digonal (sp) carbon close to 250 Hz (Table 3.14). The other major influence on J is the presence of electronegative atoms, which lead to coupling constants at the high end of the range, so that in an extreme case, chloroform has $^1J_{CH} = 209$ Hz, even though it has a tetrahedral carbon. The $^1J_{CH}$ coupling constants of tetrahedral carbon can be estimated using Eq. 3.19 and the data in Table 3.15 at the end of this chapter.

The $^1J_{CH}$ coupling constants can be measured from the ^{13}C spectrum, but they are more usually measured in the proton NMR spectrum, because fully coupled spectra like that in Fig. 3.11 are rarely taken. Most of the proton NMR spectrum is unaffected by the presence of ^{13}C: 99% of the signal from a proton is from those protons attached to ^{12}C, and these are not coupled because ^{12}C is magnetically inert; 1% of the signal, however, comes from protons attached to ^{13}C nuclei, and these are coupled, showing up—when the uncoupled proton signal is a singlet—as a weak doublet placed symmetrically about the strong signal from the protons attached to ^{12}C, and with the two lines separated from each other by an amount in Hz equal to the coupling constant $^1J_{CH}$. Since each line of this doublet is only 0.5% of the intensity of the main signal, it often goes unnoticed, but it can be searched for, if that region of the spectrum is not crowded with other signals and if the S/N ratio is good enough. All of the ^{13}C satellites, as they are called, can just be seen in

Figs. 3.5 and 3.6, and they were included, as they should be, in the integration figures quoted in connection with the text for Fig. 3.6. When the proton signal is itself a multiplet, the ^{13}C satellites are even weaker, because they are also multiplets, and they are then harder to pick out from the noise.

The coupling of a ^{13}C nucleus to a proton through more than one bond has much smaller coupling constants (Table 3.16), measurable in a full ^{13}C spectrum, when the signals are not too confused. Counter-intuitively, $^2J_{CH}$ and $^3J_{CH}$ coupling constants are rather similar, usually between 0 and 10 Hz and typically about 5 Hz. $^2J_{CH}$ values are significantly larger when the carbon is digonal (~50 Hz) or adjacent to an aldehyde group (~30 Hz), or carries an electronegative element, and significantly lower (0-3 Hz) when an alkene or arene carbon is coupled to a proton on the adjacent carbon. $^3J_{CH}$ values are affected by the dihedral angle, being large (5-10 Hz) when the angle is 180° and zero when the angle is 0°.

These longer-range couplings can be seen in expansions of the spectrum in Fig. 3.11. Beginning at the high-field end with the two quartets from the methyl groups, each line of the 1:3:3:1 quartet from the O-methyl group (at δ 59.3 in Fig. 3.11) is actually a fine 1:2:1 triplet, which can be seen in Fig. 3.13a. The carbon atom of the OMe group is three bonds away from the methylene protons on the other side of the oxygen atom, and we can detect this relationship by the $^3J_{CH}$ coupling. The coupling constant for the $^1J_{CH}$ quartet is 142 Hz, at the upper end of the range for tetrahedral carbon, because it is bonded to an oxygen atom, and the coupling constant for the $^3J_{CH}$ triplet is 5.1 Hz, a typical value. Similarly, the C-methyl groups on C-3 and C-5, which give rise to the 1:3:3:1 quartet at δ 21.2 in Fig. 3.11, actually has fine $^3J_{CH}$ coupling to the two nearly equivalent *ortho* protons (on C-2 and C-4 for the C-3 methyl group and C-6 and C-4 for the C-5 methyl group), making each line of the quartet into a 1:2:1 triplet, which can be seen in Fig. 3.13b. The coupling constant for the $^1J_{CH}$ quartet is 126 Hz and the coupling constant for the $^3J_{CH}$ triplet is 4.9 Hz. Both signals in Fig. 3.13 are called quartets of triplets.

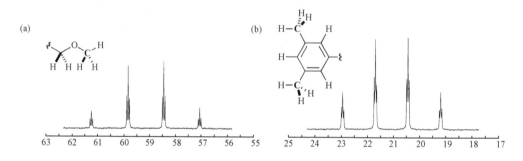

Fig. 3.13

In Fig. 3.11, the two methylene carbons are triplets close to each other. When we expand them in Fig. 3.14, we can see that each line of the downfield triplet is a fine 1:3:3:1 quartet, whereas each line of the upfield triplet is a 1:2:1 triplet. The former is described as a triplet of quartets and the latter as a triple triplet. The $^1J_{CH}$ coupling constants are 144 Hz and 147 Hz, respectively, and the $^3J_{CH}$ coupling constants are 5.2 Hz and 4.6 Hz, respectively. This pattern shows that we have assigned these two carbons correctly, because the downfield carbon is coupled ($^3J_{CH}$) to the three protons on the O-methyl group, and the upfield carbon is coupled ($^3J_{CH}$) to the two *ortho*-protons in the aromatic ring. Note that the predicted chemical shifts for these two carbons in Fig. 3.10 were only

0.9 p.p.m. apart, making it quite plausible that we could have assigned them the wrong way round.

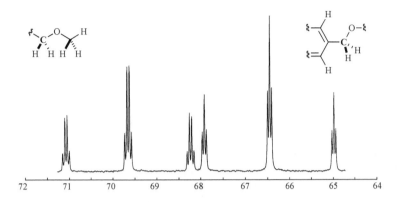

Fig. 3.14

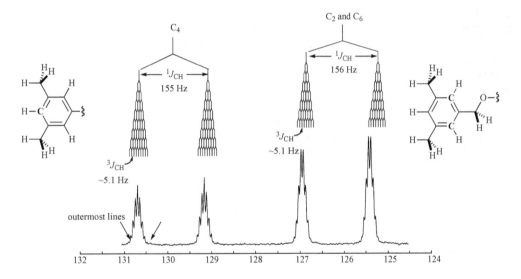

Fig. 3.15

The carbon atoms having only one hydrogen attached, C-4 and the pair C-2 and C-6, are doublets in Fig. 3.11, but the expansion in Fig. 3.15 shows that they are more complicated, each line of the doublet in the signal for C-4 is made up of nine lines, and each line of the doublet in the combined signal for C-2 and C-6 is made up of eight lines. The carbon C-4 is three bonds away from the six methyl protons and three bonds away from the protons on C-2 and C-6. The total number of hydrogens is eight, and the multiplet is therefore made up of nine lines. Similarly, C-2 (and C-6) is three bonds away from the two methylene protons, three bonds away from the protons on C-4 and C-6 (C-2), and three bonds away from the methyl group on C-3 (C-5). The total is seven protons, and the multiplet is made up of eight lines. The protons coupling to C-4 are not all identical. While the coupling constant to all six of the methyl protons will be identical, it will not be exactly the same as the coupling to the protons on C-2 and C-6. As a result, the internal lines of the nine-line pattern do not perfectly coincide, as the internal lines do for the quartets and triplets in Figs. 3.13 and 3.14. Similarly, the four different kinds of protons coupled to the carbon atoms C-2 and C-6 lead to even more broadened internal lines in the two octets. As a result, the multiplets in Fig. 3.15 are not as well resolved as they are in the two earlier figures, and the coupling constant deduced from the separation of any two lines (a little over 5 Hz in each case) is not an accurate measure of any of the component coupling constants. The only reliable number comes from the separation of the outer lines of the whole signal, which measures the sum of all the coupling constants. In the case of C-4, for example, the separation of the outer lines is 197 Hz, which must be $(^1J_{CH} + 6 \times J_{Me} + 2 \times J_{2,6})$. Also, note how small the outer lines are for the eight- and nine-line multiplets. The outer lines from the nine-line pattern within the doublet from C-4, for example, are only just discernible in Fig. 3.15, where they are picked out with arrows. A nine-line multiplet with exactly equal coupling constants would have intensity ratios of 1:8:28:56:70:56:28:8:1, and the outermost lines might easily be overlooked, just as they can be here. The rest of the signal (8:28:56:70:56:28:8 = 1:3.5:7:8.75:7:3.5:1) could be mistaken for a septet, unless one remembered that a septet would have a much steeper pattern of intensities (1:6:15:20:15:6:1).

Finally, expansion of the singlets in Fig. 3.11 shows that they are also narrow multiplets, but this time with some $^2J_{CH}$ coupling as well as $^3J_{CH}$ coupling. The expanded singlet in Fig. 3.16a from the carbonyl carbon looks similar to a quintet, which indicates that it has $^2J_{CH}$ coupling to one pair of methylene protons and $^3J_{CH}$ coupling to the other pair of methylene protons with closely similar but not quite equal coupling constants of about 4 Hz. The signal from C-3 and C-5 looks like a 1:3:3:1 quartet, indicating that it has $^2J_{CH}$ coupling of about 6 Hz to the protons on the methyl group, but undetectable $^2J_{CH}$ coupling to the protons *ortho* to it. Similarly, the signal from C-1 looks like a 1:2:1 triplet, with $^2J_{CH}$ coupling of about 4 Hz to the protons on the benzylic methylene group, but essentially zero $^2J_{CH}$ coupling to the *ortho* protons. $^2J_{CH}$ coupling to *ortho* protons is typically 1 Hz, which would not be resolved here.

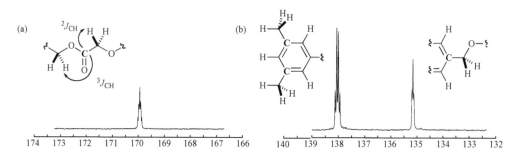

Fig. 3.16

^{13}C-^{1}H Coupling is rarely examined in the detail we have seen above, because it is rarely needed, but on this occasion it has allowed us to see the appearance of doublets, triplets, quartets, quintets and even eight- and nine-line patterns, many of which we shall see repeatedly when we come to look at the much larger and more important subject of ^{1}H-^{1}H coupling.

3.5.3 ^{13}C-^{13}C Coupling

Because of the low natural abundance of ^{13}C, it is improbable for one ^{13}C to be bonded to another. Any signals coming from such rare combinations are usually too weak to use, but enrichment with ^{13}C is now common in mechanistic and biosynthetic studies, and it is then possible to see the coupling. The geometry (tetrahedral, trigonal, digonal) is the main factor affecting $^{1}J_{CC}$. The ^{1}J coupling constants between two carbon nuclei C^x and C^y can be estimated using the expression:

$$^{1}J_{C^xC^y} = 0.073(\%s_x)(\%s_y) - 17 \tag{3.6}$$

where $\%s_x$ and $\%s_y$ are the percentages of s character (using the sp^n notation) in C^x and C^y. Thus, the (tetrahedral) methyl group in toluene is estimated to be coupled to the (trigonal) ipso carbon with a coupling constant of 43 Hz; the observed value is 44 Hz. As with $^{1}J_{CH}$, neighbouring electronegative elements raise the coupling constants; for example, the methyl group of acetates is coupled to the carbonyl carbon with ^{1}J of 59 Hz, at the upper end of the range of C—C coupling constants.

3.6 ^{1}H-^{1}H Vicinal coupling ($^{3}J_{HH}$)

We have already seen with ^{13}C-^{1}H coupling what happens when two nuclei with $I = 1/2$ are coupled. Much the same is true for proton-proton coupling, except that we are now principally concerned with two- and three-bond coupling. Two-bond coupling, often called geminal coupling, is found in methylene groups **29**, and three-bond coupling, often called vicinal coupling, is found in the arrangement **30**.

Let us begin with vicinal coupling, where the coupling constants are generally in the range 0-20 Hz. The factors that affect the coupling constant are discussed later, in Sec. 3.9, but for now we shall look only at the multiplicity. In proton NMR spectra, as with ^{13}C spectra, we see doublets, triplets, and quartets whenever a proton is coupled equally to one, two or three protons, respectively. To take a simple case, the expanded signals in Fig. 3.17 show the coupling of the aldehyde proton in diphenylacetaldehyde **31** with the proton on the α-carbon. Ignoring for the moment the confused-looking signals in the δ 7.5-7.0 region from the aromatic protons, we can see that the signal from each of the two protons is split into a doublet by the other, and the pattern is described as being that of an AX system. The convention used is to label protons close in chemical shift with the letters A, B, and C, those far away in chemical shift with the letters X, Y, Z, and those intermediate with the letters M, N and O.

Moving on to a slightly more complicated system, Fig. 3.18 shows the AX$_3$ system in the ^{1}H NMR spectrum of 2-chloropropionic acid **32**. The mid-field signal centred at δ 4.44 is the signal from the methine hydrogen H_α, downfield because it has a carbonyl group and an electronegative element attached to the methine carbon. It resonates as a 1:3:3:1 quartet because the methine hydrogen is coupled to the three identical hydrogens

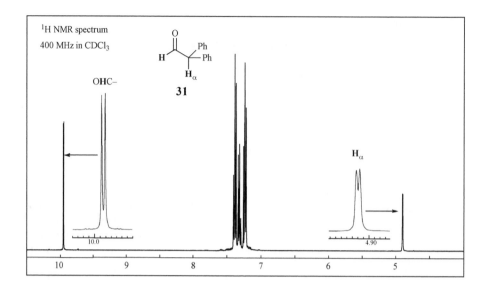

Fig. 3.17

H_β of the methyl group. Likewise, the upfield signal centred at δ 1.725 is the signal from the methyl hydrogens H_β with a chemical shift slightly downfield from the position of a normal C-Me group, because it has an electronegative element on the next carbon. It appears as a doublet because the three hydrogens of the methyl group are coupled to the single methine hydrogen, and are split into two by it. The three methyl hydrogens are, because of the free rotation about the C—C bond, identical; they experience identical magnetic environments and they come into resonance at exactly the same place. The methyl protons are, in fact, coupled to each other, but *coupling between protons with*

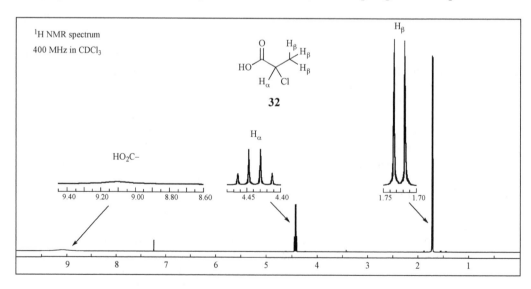

Fig. 3.18

identical chemical shifts does not show up in NMR spectra. The upfield doublet is three times as intense as the downfield quartet and three times as intense as the broad signal from the carboxylic acid proton. In summary, the mid-field signal is a one-proton *quartet* because the proton that gives rise to it is equally coupled to *three* protons, and the upfield signal is a three-proton *doublet* because the protons that give rise to it are coupled to *one* proton. The rule is the same as that given in the section on ^{13}C-^{1}H coupling: a nucleus, equally coupled to n others, will give rise to a signal with $(n + 1)$ lines, and the intensities are given by the coefficients of the terms in the expansion of $(x + 1)^n$ (Table 3.31).

The spectrum in Fig. 3.19 of ethyl propionate **33** twice over illustrates the characteristic appearance of A_2X_3 signals from ethyl groups. The three protons of the methyl groups couple equally with the two protons of the neighbouring methylene groups. Likewise, the two protons of the methylene groups couple equally with the three protons of the neighbouring methyl groups. This pattern of an upfield three-proton 1:2:1 triplet and a downfield two-proton 1:3:3:1 quartet is characteristic of an ethyl group in which the methylene protons are not coupled to anything else. The chemical shifts of the methylene groups δ 4.19 and 2.38 are strongly indicative of the nature of the atom to which they are bonded—oxygen for the former and carbon for the latter.

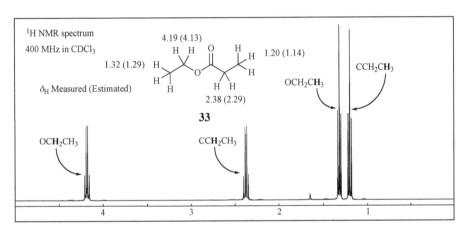

Fig. 3.19

Although it is reasonably certain that the OCH_2 signal is downfield from the CCH_2 signal, the assignment of which signal comes from which methyl group is less secure. We have assumed, correctly as it happens, that the methyl group of the OEt group will give rise to the triplet at lower field δ 1.32 than the triplet from the methyl group of the CEt group δ 1.20. The ChemNMR estimates and the measured values are shown in the inset in Fig. 3.19, where we can see that they support this expectation, but the difference in chemical shift values is not large enough for us to be completely confident. There are a number of ways in which we can confirm the assignment, as we shall see by matching coupling constants (Fig. 3.23), by difference decoupling (Sec. 3.12) and, best of all, by using COSY spectra (Sec. 3.18).

Moving on to larger multiplets, the $A_2X_2A_2$ spectrum of oxetane **34** in Fig. 3.20 shows the downfield four-proton triplet from the pair of identical methylene groups flanking the central methylene group, which gives rise to the clean 1:4:6:4:1 quintet. The triplet is downfield, because the methylene groups giving rise to this signal are adjacent to the oxygen atom. Note that the quintet here has base-line resolution and lines in the proper proportions, unlike the quintet-like signal in Fig. 3.16a. All the couplings in oxetane **34** are equal, whereas the two couplings, $^2J_{CH}$ and $^3J_{CH}$, to the carbonyl carbon in ester **1**, although similar in magnitude, were not exactly equal.

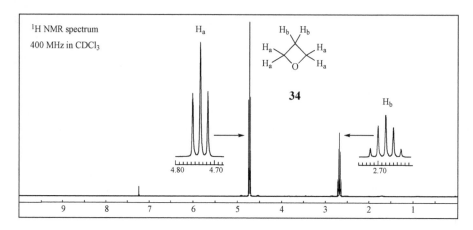

Fig. 3.20

A slightly more complicated example is shown in the spectrum of 1-nitropropane **35** in Fig. 3.21. The protons H_c on the methyl group give rise to a three-proton triplet at high field (δ 1.04), because they are adjacent to a methylene group. The protons on the methylene group H_a give rise to a two-proton triplet at low field (δ 4.37). The chemical shift is appropriate for a methylene group next to an electronegative and anisotropic group, and the multiplicity is appropriate for protons coupling to another methylene group. The protons of the methylene group in the middle H_b give rise to a two-proton 1:5:10:10:5:1 sextet at δ 2.06. The chemical shift is appropriate for a methylene group between two alkyl groups, but not far from an electronegative group. The multiplicity is appropriate for protons coupling equally to a total of five protons. Actually the coupling

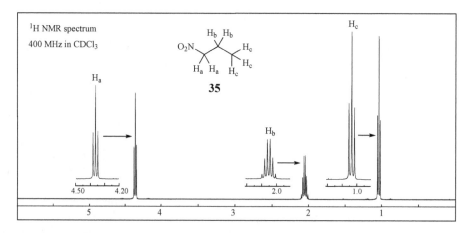

Fig. 3.21

constant J_{ab} (7.0 Hz) is slightly smaller than J_{bc} (7.5 Hz), but the difference is not resolved in the sextet, showing up only as a slight broadening of the lines and resolution that does not quite reach the base line.

In the spectra in Figs. 3.18-3.21 the coupling constants have all been very much the same, either inherently, as in the spectra of 2-chloropropionic acid **32** and ethyl propionate **33**, or accidentally, as in the spectra in oxetane **34**, where *cis* and *trans* coupling need not be equal, and nitropropane **35**, where the coupling J_{ab} is almost the same as J_{bc}. The coupling constant $^3J_{HH}$ = 6-8 Hz in these four spectra is typical of coupling constants in freely rotating alkyl chains. However, the coupling constant for the mutually coupled doublets in diphenylacetaldehyde **31** shown in Fig. 3.17 is noticeably smaller, J = 2.6 Hz. When a multiplet is split again by coupling to other protons with a different coupling constant, more complicated patterns emerge than the doublets, triplets, quartets, quintets and sextets that we have seen in Figs. 3.17-3.21.

For example, in the spectrum of propionaldehyde **36** in Fig. 3.22 the methylene protons are not the quintet that would be produced by coupling equally to the four neighbouring protons. Instead, the coupling between the methylene protons and the aldehyde proton has a coupling constant of 1.3 Hz, whereas the coupling constant to the methyl protons is 7.5 Hz. The methylene signal is therefore a double quartet, made up in the pattern shown above the expanded signal for H_α. Note how the two coupling constants can be measured in each of the participating signals, the smaller coupling constant both in the aldehyde triplet and in the double quartet, and the larger coupling constant both in the methyl triplet and in the double quartet.

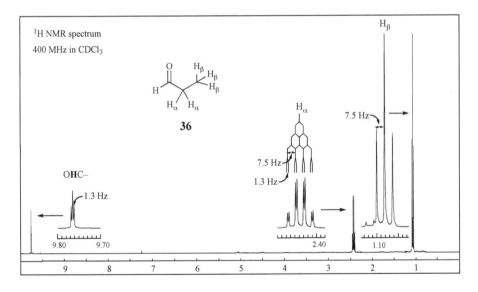

Fig. 3.22

Proton-derived signals often have to be reported in the experimental sections of research papers and in text-based compilations of data. The form in which they are reported varies with the requirements of the journal or company policy, but a typical way of reporting the spectrum of the aldehyde **36** is:

δ 9.765 (1H, t, J 1.3), 2.44 (2H, qd, J 7.5 and 1.3) and 1.08 (3H, t, J 7.5)

The order in which the chemical shift, intensity, multiplicity and coupling constant(s) are printed might vary, but a system like this is concise and easily understood. The convention here is to start at the low-field end of the spectrum (with the larger chemical shifts) and read the spectrum from left to right; the coupling constants are reported in order of decreasing magnitude, with the designations s, d, t, etc. in the same order as the coupling constants (so that the quartet above is identified as having the J value of 7.5 Hz). Occasionally, coupling constants measured on the spectrum are not completely consistent. Instruments are not immaculate in this respect, especially in the second place of decimals, since they report what the computer produces from its algorithms. The problem is common when all the coupling is not perfectly resolved. It is wise to make it clear when reporting coupling constants whether you have rationalised them (i.e. made them match up in what seems to be the obvious way) or whether you are reporting exactly what the instrument gives you. Carefully matching the coupling constants can help in the assignment of signals in complicated spectra. As a simple example, we can go back to the spectrum of ethyl propionate **33**. In Fig. 3.19, both of the coupling constants look to be about the same, but enlarging them as in Fig. 3.23 (or reading the data from the NMR spectrometer) reveals that the downfield quartet has a coupling constant of 7.14 Hz, while the upfield quartet has a coupling constant of 7.58 Hz. If we look at the two triplets, we can see that the downfield triplet has the smaller coupling constant and the upfield triplet the larger coupling constant. Being able to pair up the signals like this shows that the assignment in Fig. 3.19 was correct.

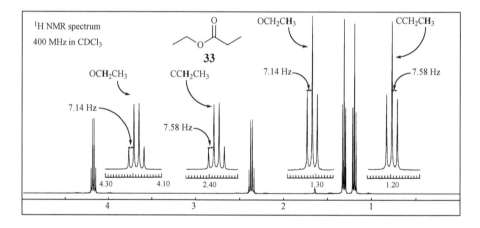

Fig. 3.23

In the double quartet in Fig. 3.22, the quartet has a large coupling constant and the doublet a much smaller one. In contrast, in the triple quartet in Fig. 3.14, the quartet has the smaller coupling constant and the triplet a much larger one. In both cases, the pattern is easy to discern—the coupling constants are so different from each other that the individual components of the signal are well separated—but in many cases individual protons give rise to patterns of lines that are much less obvious. Thus, the proton H_c on the double bond of allyl bromide **37**, centred at δ 6.03, is doubled by coupling to the *trans* proton H_a, doubled again by coupling to the *cis* proton H_b, and it is further coupled to the two protons H_d on the methylene group. Since the three coupling constants are all different, it is a double, double, triplet, and, could give rise to as many as 12 lines. The actual appearance of this signal is shown in Fig. 3.24, together with the analysis in the descending tree-like drawing above the spectrum. Because the couplings J_{cd} (7.5 Hz) and J_{ca} (10 Hz) add up to a number very close to J_{ca} (17 Hz), the two central lines almost perfectly coincide, and only 10 lines are resolved.

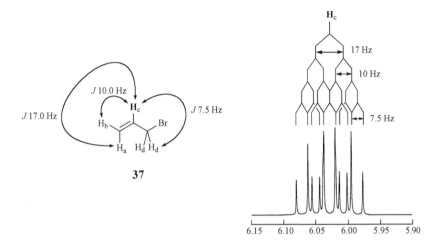

Fig. 3.24

In general, if a proton has as neighbours sets n_a, n_b, n_c... of chemically equivalent protons, the multiplicity of its resonance will be $(n_a + 1)(n_b + 1)(n_c + 1)...$, but it is not uncommon for lines to coincide, as in the example above; nor is it uncommon for protons that are not chemically equivalent to have coincidental coupling constants, as in nitropropane (Fig.3.21). In both cases, the observed number of lines is fewer than this formula suggests, and in the general case a large variety of patterns can emerge. Recognising them is a skill of great value in interpreting ^{1}H NMR spectra. It is lazy to

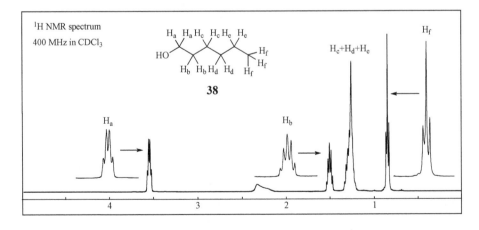

Fig. 3.25

report a signal like the one shown in Fig. 3.24 simply as a multiplet, when its components can be analysed with a little thought. In contrast, when signals seriously overlap, so that they cannot be disentangled, reporting them as multiplets is perfectly acceptable. For example, n-hexanol **38** has the spectrum in Fig. 3.25. The downfield quartet at δ 3.56 is produced by the methylene protons H_a adjacent to the oxygen atom, and the next methylene protons H_b give rise to the quintet at δ 1.51. But the remaining three methylene groups, H_c, H_d and H_e, are so similar in environment that they are not resolved. Even though the first-order analysis used here (see below for further details) predicts that they will give rise to a quintet, a quintet and a sextet, none of these patterns can be discerned, and the signal must be reported as a multiplet. The broad unresolved signal that they give rise to in the range δ 1.35-1.18 is often called a *methylene envelope*. The protons H_f of the methyl group, giving rise to the quartet at δ 0.84, resonate outside the methylene envelope, because methyl protons are usually at higher field than methylene protons.

In all the spectra considered so far, the separation of the signals (in Hz) has been much greater than the coupling constants (in Hz)—they have all been A_nX_m systems, which have allowed us to interpret the spectra using what is called *the first-order approximation*. A_nB_m systems show a deviation from first-order spectra. The simplest case is an AB system, consisting of two mutually coupled protons A and B, which are not coupled to any other protons, and which are close in chemical shift. In particular, when the difference in chemical shift between the A and the B signal ($\delta_A - \delta_B$) is comparable in magnitude to the coupling constant J_{AB}, the two lines of the doublets are not equal in intensity. As an example we can see in Fig. 3.26 the alkene signals from the mono-ethyl ester **39** of fumaric acid, in which the two protons, H_α and H_β, are in slightly different chemical environments. As a result they have slightly different chemical shifts (δ 6.97 and 6.87). The difference in chemical shift is 40 Hz and the coupling constant of 17 Hz is not very different. In consequence, the 'inside' lines of the AB system are more intense, and the 'outside' lines less intense. The comparison is with a simple, first-order AX system, where the lines in the A and the X signals are essentially equal in intensity (Fig. 3.17).

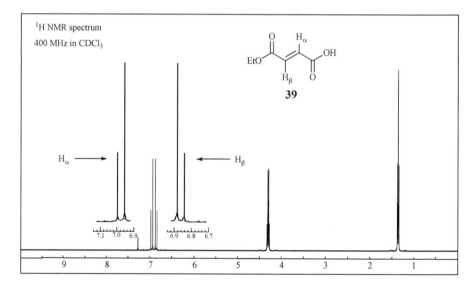

Fig. 3.26

The formulae governing the appearance of AB patterns are given in Eqs. 3.7 to 3.9 , with the symbols explained in Fig. 3.27. The smaller the chemical shift difference, the more the inside lines, 3 and 2, grow in size, and the outside lines, 4 and 1, get smaller. The relative intensity of the lines, I, is given by:

$$\frac{I_3}{I_4} = \frac{I_2}{I_1} = \frac{(v_4 - v_1)}{(v_3 - v_2)} \tag{3.7}$$

The coupling constants are given in the same way as in an AX system:

$$J_{AB} = v_4 - v_3 = v_2 - v_1 \tag{3.8}$$

In an AX system, the chemical shifts of the A and the X signals are given by the frequency of the midpoints of each of the doublets. This is no longer the case with an AB system, where the chemical shifts δ_A and δ_B are given by:

$$\delta_A - \delta_B = \sqrt{(v_4 - v_1)(v_3 - v_2)} \tag{3.9}$$

As illustrated in Fig. 3.27c, this places the true chemical shifts closer to the inside lines than to the outside lines.

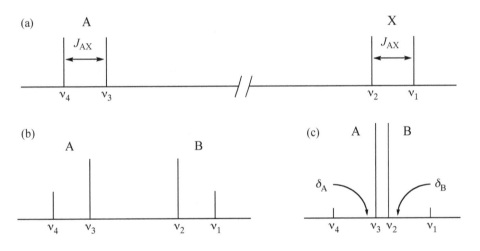

Fig. 3.27

Thus, the closer the signals are in chemical shift, the greater the perturbation, as we can see by comparing the AB systems in Fig. 3.28 given by the protons on the double bonds in three α,β-unsaturated carbonyl compounds. The protons in methyl 3-methoxyacrylate **40** have very different chemical shifts, because the β-proton is adjacent to a σ-withdrawing substituent, conjugated, and *cis* to a π-withdrawing substituent; in contrast, the α-proton is conjugated to a π-donor substituent (see Table 3.3 and the associated text for a revision of these points). The values are δ 7.65 and 5.21, respectively, more or less at the extreme range for olefinic protons. At 400 MHz, this is a difference of 976 Hz, and the coupling constant is much smaller at 13 Hz. As a result, the signals are only slightly perturbed from those of an AX system. Santonin **41** has the two signals separated by a much smaller amount, 172 Hz at 400 MHz, with a somewhat smaller coupling constant, 10 Hz, and the perturbation is more obvious. The ester **39** that we have already seen has a separation in chemical shift of only 40 Hz in the 400 MHz spectrum with a large coupling constant of 17 Hz, and the perturbation is considerable.

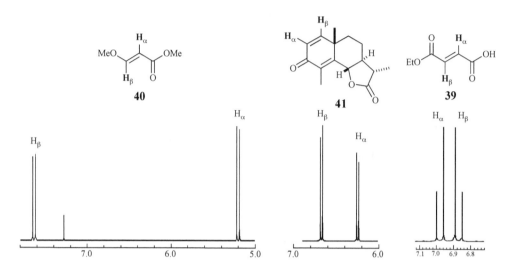

Fig. 3.28

These patterns are helpful in identifying AB systems: a strongly perturbed doublet must be coupling to a proton close in chemical shift and a less perturbed doublet to one further away. Furthermore, the perturbation tells us in which direction to look for the other half of the AB system. Nevertheless, it is always wise when you are assigning signals to measure up the doublets, in order to make sure that both halves of what you think are an AB system have matching coupling constants.

The doublet of each partner in an AB system is described as 'pointing' to its partner. Alternatively the AB system as a whole is described as 'roofing', where the metaphor comes from the picture in Fig. 3.27b with the 'roof' drawn from the top of the outside peaks to the top of their neighbours and continuing up to, and meeting, at the 'roofline' at the centre of the system. At the extreme, when A and B have exactly the same chemical shift, the outside lines disappear, and the inside lines merge into a singlet. This is the situation with the methylene groups in the compounds **33-38** used in Figs 3.19-3.25. In these cases, the methylene hydrogens are *inherently* identical, but coupling also disappears when two chemically distinct protons *accidentally* come into resonance at the same frequency. The inside lines merge and the outside lines disappear. When the inside lines do not quite merge, the outside lines can be so small as to be overlooked, and the signal can easily be mistaken for a doublet.

The same type of perturbation occurs in all A_nB_m systems, and will be evident in many of the spectra in this chapter. Looking back, it is even possible, just, to see the effect in such spectra as that of ethyl propionate **33** in Fig. 3.23, where the triplets of both A_2X_3 systems point to the quartets, with the triplet from the C-Et group, closer in chemical shift to its partner, pointing a little more strongly than the triplet from the O-Et group, further apart in chemical shift from its partner.

Roofing (or pointing) can be helpful in making sense of a complex pattern, such as that from the three olefinic protons in methyl acrylate **42** shown in Fig. 3.29.

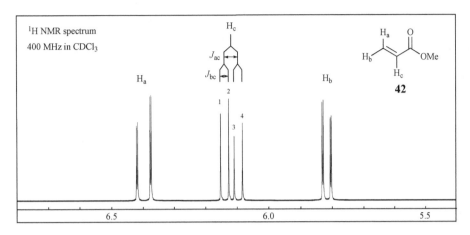

Fig. 3.29

We can expect all three protons to be downfield (Table 3.3), and the *cis*-β-proton H_a to be the most downfield of all. Judging by methyl vinyl ketone **17** (Table 3.3), the α-proton H_c will be in between the two β-protons, as indeed it is. This assignment is reinforced by looking at the signal from the α-proton, which is a straightforward double doublet with coupling constants of 17.2 and 10.8 Hz. The larger separation of the lines 1 and 3, and the equally large separation of the lines 2 and 4, gives the coupling constant J_{ac}. This is confirmed by the roofing, since both these pairs point towards the downfield signal from H_a, which also has the larger coupling constant. Similarly, the smaller separation of the lines 1 and 2, and the equally small separation of the lines 3 and 4, gives the coupling constant J_{bc}, and this too is confirmed by the roofing, since both pairs point towards the upfield signal from H_b. In turn, the doublets given by H_a and H_b point back to the central signal given by H_c. Of course the same assignments can be made simply by looking at the coupling constants, but the roofing is a great help in quickly making sense of the appearance of the signal from H_c.

One should be cautious in assigning coupling constants using only the first-order analysis given here, because the separation of, say, lines 1 and 2 and lines 1 and 3 is not an accurate measure of the two coupling constants. More often than not coupling constants are reported by measuring these separations, because J_{bc} is close to the separation of lines 1 and 2, and J_{ac} is close to the separation of lines 1 and 3. Nevertheless, it is strictly true only that the separation of lines 1 and 4 is the sum of J_{bc} and J_{ac}. This warning applies to all multiple spin systems, but we shall continue to use the first-order simplification, because the differences from the true coupling constants are not normally significant.

Returning to the spectrum in Fig. 3.29 we might note that the signals from both H_a and H_b are actually double doublets with fine coupling to each other. This is the first example of $^2J_{HH}$ coupling that we have seen, and it takes us to the next section.

3.7 ^{1}H-^{1}H Geminal coupling ($^2J_{HH}$)

Geminal coupling $^2J_{HH}$, also known as two-bond coupling, is found only in methylene groups **29** in which for some reason the two hydrogens H$_A$ and H$_B$ are not identical and do not therefore come into resonance at the same frequency. They give rise to multiplets, in the first-order approximation, with the same rules as for three-bond coupling, and the range of coupling constants is rather similar, 0-25 Hz. Not surprisingly, two hydrogens bonded to the same carbon atom are frequently close in chemical shift, and roofing is almost always visible.

The two hydrogens of a methylene group are different in terminal alkenes, as we have seen in methyl acrylate **42**, where they split each other with a coupling constant of only 1.2 Hz. They are also different in cyclic compounds when one surface of the ring has different substituents from the other surface, as in the epoxide **43**, which has an aromatic acyloxymethyl group on one surface and a methyl group on the other. The methylene protons, H$_a$ and H$_b$, are in different environments, and they give rise to an AB system with another small coupling constant, 2.6 Hz, and an appropriately small amount of roofing (Fig. 3.30).

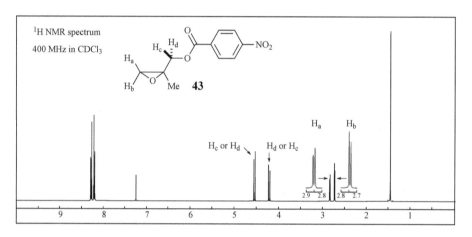

Fig. 3.30

It is not quite so obvious why, but the methylene group in the side chain, CH$_c$H$_d$, also gives rise to an AB system with a coupling constant of 12.2 Hz, large enough to be visible in Fig. 3.30 without expanding the signal, and with the two signals actually further apart than the more obviously different pair H$_a$ and H$_b$. At first sight, you might expect the two protons to be chemically identical, especially when you allow for the free rotation of the side chain, but the presence of a stereogenic centre in the molecule has the effect of placing H$_c$ and H$_d$ in different environments. There are three conformations **44-46** in which all the groups are staggered about the bond connecting the methylene group to the ring. In the first place, the side chain will probably adopt one conformation in preference to any other, and in that conformation, say **44**, the two protons H$_c$ and H$_d$ are not in the same environment, and can come into resonance at different frequencies. Secondly, even if the rotation is completely free, and all three conformations are occupied, the average field experienced by H$_c$ is not inherently the same as that experienced by H$_d$. At any one moment, say **44**, when H$_c$ is in the top left segment, H$_d$ will be placed between the epoxide methylene group and the methyl group, but when H_d comes to the top left **45**, H_c will be placed between the epoxide oxygen and the methyl group. Thus, H_d does not experience the same environment as H_c experienced when it was in the top left segment. The two conformations **44** and **45** are not identical, nor are they enantiomers. At no stage in any of the conformations **44-46** is either of the protons in the same environment as that which the other experiences as the side chain rotates. The two protons H_c and H_d are said to be *diastereotopic*. Only when the average field is the same by coincidence do diastereotopic protons, and diastereotopic methyl groups likewise, come into resonance at the same frequency. The test for diastereotopic groups is to identify what kind of stereoisomers would be produced if first one of them were replaced by a completely different group, and then the other. If H_c were replaced by a methyl group, it would create a diastereoisomer of the compound produced by changing H_d into a methyl group.

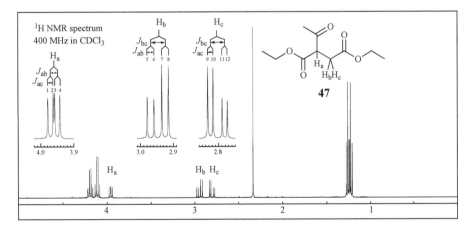

Geminal coupling commonly occurs alongside vicinal coupling, and ABX patterns are often the result. They have a wide variety of appearances, depending upon the relative chemical shifts and coupling constants of the three protons. The appearance of the ABX system in diethyl acetylsuccinate **47** is only one of many possible patterns. If we combine the treatment of an AB system from the previous section with a simple first-order prediction, we can expect that the four AB lines from H_b and H_c will each be split into doublets by coupling to the X proton H_a, as observed in Fig. 3.31. The proton H_a, being chemically well shifted from H_b and H_c, but coupled to both with different coupling constants, appears as a double doublet. The AB lines exhibit the differences in intensity stemming from their AB coupling, but the double doublet of the more distant X proton H_a has four lines nearly equal in intensity. This example shows the full total of 12 lines, since the signals are well spaced, and the coupling constants are all different: 8.3 (J_{ab}), 6.3

Fig. 3.31

(J_{ac}) and 17.7 Hz (J_{bc}). There are many other possible patterns, depending upon whether any of the lines accidentally coincides with any other, which in turn depends upon how large the separation of the A and B signals is, and what the coupling constants are. Thus, if H_b and H_c were closer in chemical shift, line 9 could easily be coincident with, or even at lower field than, line 8 or line 7. Equally, the X signal could be upfield of the AB system rather than downfield, and it could in its turn be coupled on to other protons extending the spin system beyond that of an ABX.

Separate signals from diastereotopic protons are common for a methylene group adjacent to a stereogenic centre, as in the compounds **43** and **47**, but it is even quite commonly observed when the stereogenic centre is further away, and it is not even necessary for there to be a stereogenic centre. The two ethoxy groups in the diethylacetal **48** are *enantiotopic*—replacing one with another group would create the enantiomer that would be created by replacing the other ethoxy group. But the methylene hydrogens within the ethoxy groups are diastereotopic, even though the molecule is achiral. Taking one of the hydrogens, say H_a from the front ethoxy group, and replacing it with another group would create one diastereoisomer; replacing the other hydrogen H_b would create a different diastereoisomer. The same phenomenon is equally true of the enantiotopic ethoxy group at the rear, and the two diastereoisomers this time will be enantiomers of the first two. Thus, H_a and H_b are neither chemically nor magnetically equivalent, but the two H_as are chemically and magnetically equivalent, and the two H_bs likewise. Ethoxy groups in diethyl acetals, and in chiral esters like **47**, are often too far from the source of the asymmetry—the diastereotopic protons come into resonance at the same frequency, and appear as a quartet in the usual way for an ethoxy group. In this case, however, they give rise to a complicated, highly symmetrical, but understandable set of signals in the range δ 3.85-3.68, expanded in Fig. 3.32. It consists of a pair of mutually coupled double quartets at δ 3.81 and δ 3.73, with strong roofing within the doublet component (J_{ab} of 9.6 Hz), and a smaller coupling constant ($J_{aMe} = J_{bMe} = 7.2$ Hz) for the coupling to the methyl group. As in the ABX system in Fig. 3.31, these coupling constants are approximate, but the separation of the outside lines is accurately $3J_{aMe} + J_{ab}$.

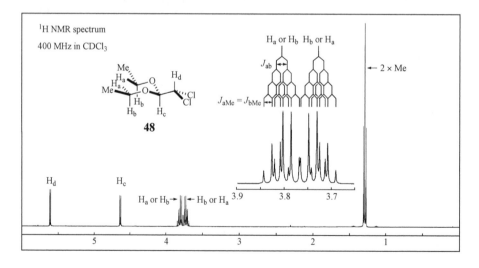

Fig. 3.32

3.8 ^{1}H-^{1}H Long-range coupling ($^4J_{HH}$ and $^5J_{HH}$)

Coupling through four or more bonds is often called long-range coupling. The coupling constants are naturally quite small, rarely outside the range 0-3 Hz. They are at the higher end of this range in two quite commonly encountered situations. The first is in unsaturated systems, when a double bond is oriented so that its π-system overlaps with a C—H σ-bond, as in the allyl, allene and propargyl systems **49-51**. The difference between the two allylic couplings shown in **49** is too small to permit a reliable assignment of geometry. Homoallylic coupling ($^5J_{HH}$ = 1-2 Hz) is sometimes resolved, but only when the allylic C—H bond overlaps with the double bond, as in the allyl and allene partial structures **52** and **53**, and is especially strong when the C—H bonds are rigidly held and doubly conjugated as in 1,4-cyclohexadienes **54**.

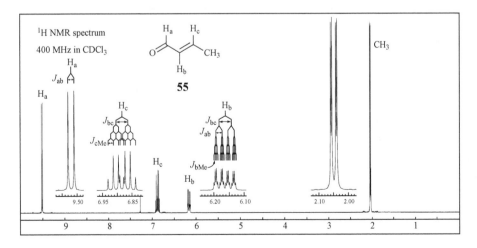

We can see an example of allylic coupling in the signal at δ 6.16 from the α-proton H_b of crotonaldehyde **55** in Fig. 3.33. This proton is vicinally coupled ($^3J_{bc}$) to the β-proton H_c with a coupling constant of 15.6 Hz; it is also coupled ($^3J_{ab}$) to the aldehyde proton H_a

Fig. 3.33

with a coupling constant of 7.9 Hz, nearly half as large. As a result, the α-proton H_b is a double doublet, consisting of four nearly equally spaced signals, but it is also allylically coupled ($^4J_{bMe}$) to each of the three protons of the methyl group, making each line of the double doublet a fine quartet with a coupling constant of 1.6 Hz. This signal, with resolved coupling from all of the other protons in the molecule, can be described as: δ 6.16 (1H, ddq, J 15.6, 7.9 and 1.6 Hz). The remaining signals match up: the aldehyde proton H_a at δ 9.52 is a doublet, with necessarily the same ^3J vicinal coupling of 7.9 Hz to the α-proton; the signal from the β-proton at δ 6.89, downfield from the α-proton, is a double quartet with coupling constants of 15.6 and 6.9 Hz; and the signal from the methyl group at δ 2.05 is a double doublet, with matching coupling of 6.9 Hz to the β-proton and allylic coupling of 1.6 Hz to the α-proton H_b.

The second commonly encountered case of long-range coupling is in rigid saturated systems. Four-bond coupling is often resolved when the four bonds adopt a planar W arrangement, as emphasised for the 1,3-diequatorial protons in rigid cyclohexanes **56** and in bicyclo[2.2.1]heptanes **57**. Again, there are exceptionally high values when the overlap of the σ-bonds is especially favourable, as in bicyclo[2.1.1]hexane **58**. W coupling is also evident in unsaturated systems, as in the frequently resolved meta coupling in aromatic rings **59**, but five-bond para coupling is rarely resolved.

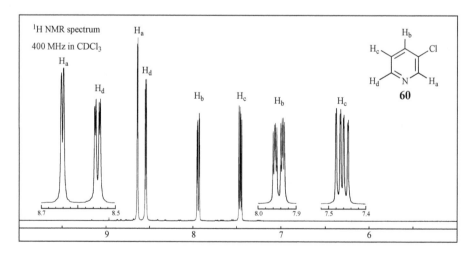

We can see long-range coupling in an aromatic system in the spectrum of 3-chloropyridine **60** in Fig. 3.34. If we ignore the fine coupling, vicinal coupling leads H_a to be a singlet, H_b to be a doublet (^3J = 8.3 Hz), H_c to be a double doublet (^3J = 8.3 and

Fig. 3.34

4.6 Hz) and H_d to be a doublet (^3J = 4.6 Hz). But each of these signals is split by meta coupling: H_a by H_b (^4J = 2.4 Hz), H_b by both H_a and H_d (^4J = 2.4 and 1.5 Hz), and H_d by H_b (^4J = 1.5 Hz). Even H_c shows some barely resolved para coupling ($^5J \approx$ 0.7 Hz), although it is not resolved in H_a.

If long-range coupling is present but not resolved, it leads simply to line broadening. Two or three of the earlier spectra used in this chapter show this phenomenon—look at the spectrum of the ester **1** in Fig. 3.6: the 3-proton line from the methoxy group is actually taller than the 6-proton line from the two aromatic methyl groups. The integral, of course, reveals that the areas under the signals are in the proportion 3:6; but if the signals were expanded we would see that the width at half height of the *C*—Me signal was greater than that of the signal from the methoxy protons. The *C*-Me signal is broadened because the protons are coupled to the protons on C-2 and C-4 by ^4J coupling similar to allylic coupling. Similarly, we can now see why the doublet from H_α in diphenylacetaldehyde **31** in Fig. 3.17 is so broad, whereas the aldehyde proton gives rise to a sharp doublet—H_α is weakly coupled to the four *ortho* protons in the aromatic rings.

As NMR instruments get better and better, resolved long-range coupling more and more often intrudes into the simple analysis. An unexpected example can be seen in the spectrum of pantolactone **61** in Fig. 3.35. The AB system of H_a and H_b with a coupling constant of 9.1 Hz is clear (δ 4.05 and 3.90). The doublet from H_b is notably shorter and broader than that of H_a, as seen in the upper expansion. Similarly, the upfield singlet from one of the two methyl groups is shorter and broader than the downfield singlet; H_b is obviously coupled to Me_a. The upper expansion of the H_b signal in Fig. 3.35 is from a spectrum taken in a routine way, but the lower expansion from another spectrum, optimised both in the taking and the processing, actually reveals the quartet structure within each line of the doublet, and a ^4J coupling constant of 0.6 Hz, matched in the expansion of the signal from Me_a. The most populated conformation of this molecule must have H_b held in a good W-arrangement with one of the hydrogen atoms of the methyl group *trans* to it Me_a. Since the methyl group is freely rotating, the observed coupling constant will have been reduced from the maximum value, because only one of the three hydrogen atoms can be in a W-arrangement at any one time, making the resolution of this signal all the more remarkable.

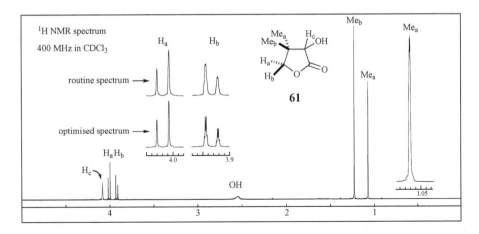

Fig. 3.35

We can also see long-range coupling again in what we called the 'confused-looking signals in the δ 7.5-7.0 region' from the aromatic protons in diphenylacetaldehyde **31** in Fig. 3.17. Monosubstituted aromatic rings like this often have overlapping signals, especially when the substituent is effectively an alkyl group. Indeed, the ChemDraw®-predicted chemical shifts for this compound are in the right order: *meta*: δ 7.33, *para*: δ 7.26, and *ortho*: δ 7.23, which would probably not be fully resolved. The effect through space from the anisotropy both of an aldehyde group and another phenyl ring has shifted the signals in ways that the program did not handle perfectly, and the three signals are in fact a little more spread out: *meta*: δ 7.41, *para*: δ 7.34, and *ortho*: δ 7.26. We are now in a position to make some sense of these signals, which are expanded in Fig. 3.36. Vicinal coupling will lead the *ortho* protons to give rise to doublets, the *meta* protons either to double doublets or to triplets (depending upon whether the coupling constants are equal or not), and the *para* protons, which will be half as intense, to triplets. Whereas H_c in the pyridine **55** had two very different *ortho* coupling constants, the more symmetrical benzene ring usually has the two coupling constants equal, and so they are in the aldehyde **31** with both $^3J_{om}$ and $^3J_{mp}$ approximately 7 Hz. But in addition to the *ortho* coupling, these signals show fine structure that stems from *meta* and maybe some *para* coupling. Thus, each of the lines in the strongly roofed 2-proton triplet in the middle from the *para* protons is split again into fine triplets, with $^4J_{op}$ of approximately 2 Hz, because of coupling to the *ortho* protons. The fine structure within the 4-proton signal from the *meta* protons may stem from para coupling, since the *meta* protons are coupled to the *ortho* in two ways, vicinal and long-range para. It may also stem from imperfect matching of the coupling constants to the *ortho* and *para* protons, and yet another explanation for the extra lines is covered in the next section.

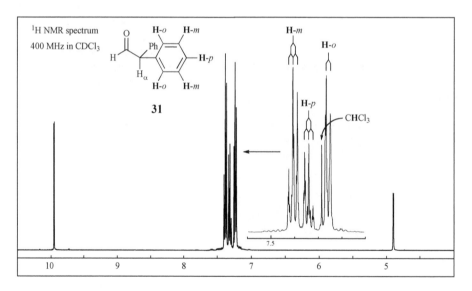

Fig. 3.36

3.9 Deviations from first-order coupling

We have not, so far, strayed far from the first-order approximation; we have only added the roofing that results when the chemical shift difference and the coupling constant are similar, and added a cautionary word that the position of the lines may not strictly allow you to measure the coupling constants. There are, however, more substantial failures of the first-order approximation when several spins are involved, and we may have begun to see this in the spectrum in Fig. 3.36. In a many proton spin system, especially one with duplicate protons like the *ortho* and *meta* protons in each aromatic ring in the aldehyde **31**, the first-order analysis is not always adequate. It is usually possible to discern the essential pattern—two triplets and a doublet with strong and appropriate roofing in this case—but extra lines are not at all uncommon. They stem from the many energy levels populated in a multi-proton system, and the many transitions not taken into account in the first-order analysis. (For some discussion of energy levels, see the next section.) A proper theoretical treatment does account for these patterns, and for those described below, but it is no longer a first-order analysis. For now, we need only know that some splitting patterns are not readily analysed just by inspection and first-order analysis. The three examples in Fig. 3.37 will suffice to illustrate the kinds of patterns that can turn up.

Fig. 3.37a shows the characteristic pattern often given by *para*-disubstituted benzenes. The aromatic protons in *p*-bromophenetole **62** might be expected to show an AB pattern, and they more or less do, but there are extra lines, conspicuously inside each of the doublets, as a consequence of there being two identical *ortho* and two identical *meta* protons. This is one type of an AA′BB′ system, and another is that shown in Fig. 3.37b, given by 1,4-diphenylbutadiene **63**, which is very similar. 3,3-Dimethylbutylamine **64** has a pair of adjacent methylene groups with no further coupling, which might have been expected to give a pair of 1:2:1 triplets, as such systems frequently do. However, the pattern produced in this case, shown in Fig. 3.37c, is clearly more complicated, although it does bear some resemblance to a pair of triplets, but with the central line broken up. (The sloping line at the downfield end of the H_b signal is the shoulder of the broad signal of the NH_2 protons.)

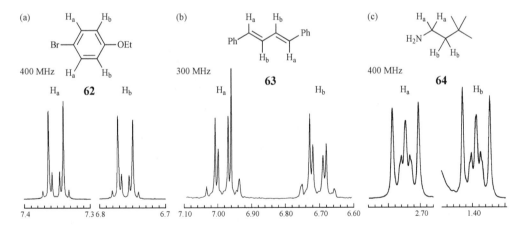

Fig. 3.37

3.10 The magnitude of ^1H-^1H coupling constants

In the course of the discussion so far, we have seen coupling constants as widely different as 0.6 Hz and 17.7 Hz. We now consider what factors most significantly affect the size of *J*. Information about the magnetic orientation of one nucleus is transmitted to the other by the intervening electrons. Transmission of information is dependent upon how well the orbitals containing those electrons overlap, as well as by the number of intervening orbitals. In a crude approximation, the number of intervening orbital interactions affects both the sign and the magnitude of the coupling constant.

Coupling constants can be either positive or negative. Although this does not affect the appearance of the spectrum, it does change the way in which structural variations affect the magnitude of the coupling constant. To understand why coupling constants can be positive or negative, we need to look a bit more carefully into the energetics of coupling. In hydrogen itself, H_2, there are three arrangements with different energies: the lowest energy with both nuclei H and H' aligned, the highest with both opposed, and in between two ways equal in energy with the alignments opposite to each other (Fig. 3.38a, where upward-pointing arrows indicate nuclear magnets in their low-energy orientation with respect to the applied magnetic field, downward-pointing arrows indicate nuclear magnets in their high-energy orientation with respect to the magnetic field, and levels of higher energy are indicated by vertical upward displacement). The transitions which the instrument measures are those in which the alignment of one of the nuclei changes from the N_β to the N_α state of Fig. 3.1. There are four such transitions labelled W in Fig. 3.38a, and all of them equal in energy difference. The receiving coils detect only the one signal, and the print-out shows one line and no coupling.

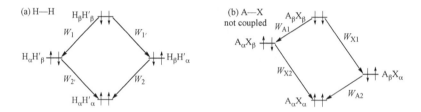

Fig. 3.38

If now we look at two different atoms A and X, we have the same set-up, but this time the two energy levels in the middle are of different energy, one with A aligned and the other with X aligned (Fig. 3.38b). 'A' might be a ^{13}C, and 'X' a ^1H atom, but the general picture is the same for all AX systems. If there is no coupling ($J = 0$), as when the nuclei are far apart, the $A_\alpha X_\beta$ energy level will be as much above the mid-point as the energy level for the $A_\beta X_\alpha$ nucleus is below it. There will again be four transitions, two equal for the A nucleus, labelled W_A, and two equal for the X nucleus, labelled W_X, giving rise to one line from each.

If, on the other hand, the two nuclei are directly bonded, they will affect each other. The A spin will be opposed to the spin of one of the intervening electrons in an s-orbital (only s-orbitals have an electron population at the nucleus); that electron is paired with the other bonding s-electron. In the lowest energy arrangement of the system, both the A and X nuclei are spin-paired with the bonding electrons with which they interact most strongly (as in the picture on the right-hand side of Fig. 3.39). As a result, the A and the X nuclei will be opposed in the lowest energy arrangement. Conversely, the system will be higher in energy when these spins are aligned. Thus, the two energy levels in which the A and X nuclei have parallel spins will be raised and the two energy levels in which they are opposed will be lowered (Fig. 3.39). Thus, there are now four new energy levels, four different transitions, W_{A1} and W_{A2}, and W_{X1} and W_{X2}, and four lines in the AX spectrum. The A signal is a doublet and the X signal is a doublet, with the same separation between the lines, because $(W_{A1} - W_{A2}) = (W_{X1} - W_{X2}) = J_{AX}$. Thus, the extent of the raising and lowering of each of the energy levels is $J_{AX}/4$. More complicated versions of this kind of diagram are needed to analyse spin interactions beyond the AX system, and even more complicated ones to make sense of those spectra that are not first order.

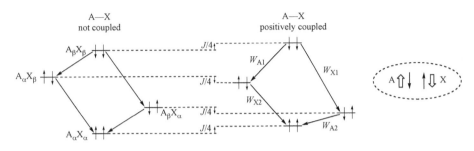

Fig. 3.39

If instead of being directly bonded, the A and X nuclei are separated by two bonds, the transmission of information through the s-electrons leads the two nuclei to be parallel in the *low*-energy arrangement, in contrast to the *high*-energy arrangement of Fig. 3.39. The model that illustrates this point is given on the right-hand side of Fig. 3.40, and implies that the nuclei will be anti-parallel in the *high*-energy arrangement. Now the energy levels will have the lowest and highest energy levels lowered by the interaction of the two spins, and the levels in between raised. If the coupling constant is the same as that in Fig. 3.39, the two transitions for the A nucleus, W_{A1} and W_{A2}, are of the same magnitude as before but have changed places, and similarly for W_{X1} and W_{X2}. The appearance of the spectrum will not have changed, but the coupling constant J is negative in sign.

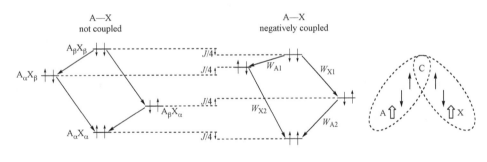

Fig. 3.40

In general, although not always, one-bond couplings ^1J and three-bond couplings ^3J are positive in sign, and two- and four-bond couplings ^2J and ^4J are negative in sign. Given this understanding, we now discuss separately the factors affecting coupling constants for two-, three- and four-bond coupling.

3.10.1 Vicinal coupling $^3J_{HH}$

The dihedral angle. Coupling is mediated by the interaction of orbitals within the bonding framework. It is therefore dependent upon overlap of the orbitals, and hence upon the dihedral angle between the bonds that are involved. The relationship between the dihedral angle and the vicinal coupling constant ^3J is given theoretically by the Karplus equations:

$$^3J_{ab} = J_0 \cos^2 \phi - 0.28 \quad (0° \leq \phi \leq 90°) \tag{3.10}$$

$$^3J_{ab} = J_{180} \cos^2 \phi - 0.28 \quad (90° \leq \phi \leq 180°) \tag{3.11}$$

where J_0 and J_{180} are constants which depend upon the substituents on the carbon atoms and ϕ is the dihedral angle defined in Fig. 3.41.

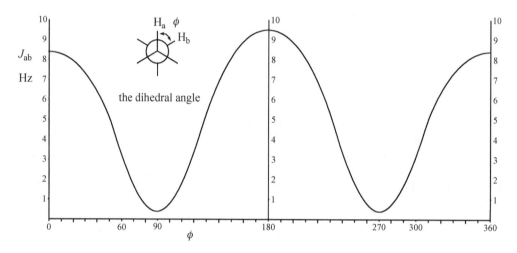

Fig. 3.41

The Karplus equations are plotted in Fig. 3.41 using $J_0 = 8.5$ and $J_{180} = 9.5$, the standard values when no better estimate is available. Coupling constants observed experimentally follow this relationship well, but it is not always easy to choose values of J_0 and J_{180}. The main point to notice is that the coupling constant is at its largest when the dihedral angle is 180°, in other words, when the hydrogens are antiperiplanar and the orbitals are overlapping most efficiently; slightly smaller when it is 0°, when they are syncoplanar; and at its lowest when the dihedral angle is 90° and the orbitals are orthogonal. In an ethyl group, the free rotation allows the vicinal hydrogens to pass through all these angles, but they will spend most of their time in the usual staggered conformation, with dihedral angles of 60°, 180° and 300°. The coupling constants for an ethyl group that we have seen in Figs 3.21, 3.22, 3.23, 3.25 and 3.32 are all close to 7 Hz, which is near the average of the coupling constants given by the Karplus equation for these three angles.

In rigid systems, where averaging is not possible, we frequently get both larger and smaller values. In rigid cyclohexanes **65**, for example, the axial-axial coupling constant, J_{aa}, is usually large, in the range 9-13 Hz, because the dihedral angle is close to 180°. The axial-equatorial and equatorial-equatorial coupling constants, J_{ae}, and J_{ee}, are much smaller, usually in the range 2-5 Hz, because the dihedral angles are close to 60°. The dihedral angles are clearer on the Newman projections **66** and **67**, but it should be remembered that the bond angles are not always so perfectly bisected in real systems.

Nevertheless, these differences in a well-behaved and relatively rigid system are large enough to make this a powerful tool in assigning stereochemistry. The conformations of cyclopentanes are much less predictable—*cis* coupling constants are sometimes higher and sometimes lower than *trans* coupling constants—making stereochemical assignments based on coupling unreliable in 5-membered rings.

We can now see why the vicinal coupling constants in the ABX system in Fig. 3.31 are different. The acetylsuccinate will mainly adopt the conformations **68** and **69** with the carbonyl groups as far apart as possible, with the third gauche conformation **70** clearly less favourable. In the conformation **68**, H_a and H_b have a dihedral angle of 180° and hence a large coupling constant, and H_a and H_c have a dihedral angle of 60° and a small coupling constant. In the alternative conformation **69**, these relationships are inverted. As long as one of these two conformations is more populated than the other, the coupling constants will be different. That they have similar values, 8.3 and 6.3 Hz, indicates that both conformations are populated, but not quite equally.

A modified Karplus equation can be applied to vicinal coupling in alkenes; the numbers are slightly different, but the conclusion is the same. A dihedral angle of 180° is found in *trans* double bonds **71**, where the coupling constants are large, and a dihedral angle of 0° is found in *cis* double bonds **72**, where the coupling constants are smaller. This is exemplified in the coupling constants of 17.2 and 10.8 Hz found in methyl acrylate **42** in Fig. 3.29.

The presence of electronegative or electropositive elements. An electronegative element directly attached to the same carbon atom as one of the vicinally coupled protons reduces the coupling constant, because it reduces the electron population responsible for transmitting the coupling information. Electropositive elements raise the coupling constant. For freely rotating chains, the effect is small **73-75**. The effect of electronegative elements is cumulative, as we can see in the spectrum of the acetal **48** in Fig. 3.32, where the coupling constant between H_c and H_d, with two electronegative substituents each, has dropped to 5.2 Hz in spite of the likelihood that the two protons are held antiperiplanar, whereas those in the structures **73-75** are not.

The presence of the oxygen atom in the aldehyde group explains the small coupling constant to the neighbouring hydrogens in aldehydes like diphenylacetaldehyde **31** and propionaldehyde **36** in Figs. 3.17 and 3.22, where the coupling constants were only 2.5 and 1.3 Hz, respectively. In crotonaldehyde **55** in Fig. 3.33, on the other hand, the coupling constant is 7.9 Hz, because the effect of the electronegative element is somewhat offset by the coplanarity of the conjugated system that keeps the aldehyde proton H_a and the α-proton H_b antiperiplanar most of the time. When the electronegative element is held rigidly antiperiplanar with respect to one of the protons (heavy outline in **76**), then the effect is larger. Thus, J_{ae} is only 2.5 ± 1 Hz when X (OH, OAc or Br) is axial, but it is 5.5 ± 1 when it is equatorial **77**, even though the dihedral angles are close to 60° in both cases. On double bonds, both the antiperiplanar and the syncoplanar protons are affected. An electronegative element substantially lowers both the *cis* and the *trans* coupling constants in vinyl fluoride **78** relative to propene **79**, and an electropositive element has the opposite effect. The *cis* coupling constant for vinyl-lithium (**80**, X = Li) is higher even than the normal value for a *trans* double bond, and the *trans* coupling is higher still.

We can see an extreme example of the lowering of the coupling constant by an electronegative element in the spectrum in Fig. 3.42. 3,4-Epoxytetrahydrofuran **81** has a plane of symmetry, and therefore shows only three resonances: an AB system from the methylene protons H_a and H_b at δ 4.04 and 3.68(not necessarily respectively), with a coupling constant of 10.4 Hz; and a sharp singlet in between at δ 3.81 from the methine proton H_c. The dihedral angle between H_a and H_c is close to 90°, and it is not surprising that there is no coupling between them. However, the dihedral angle between H_b and H_c is somewhere between 0° and 30°, and a coupling constant of 6-8 Hz might be expected on the basis of the Karplus equation, whereas in fact the only sign of coupling is the slight broadness of the upfield signal. A major influence is the epoxide oxygen anti to proton H_b, but angle strain also contributes to the disappearance of coupling.

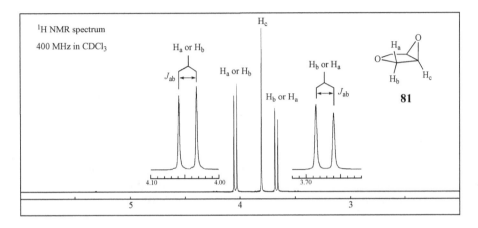

Fig. 3.42

Angle strain. In the fragment **82**, orbital overlap, and hence ^3J, decreases as θ and θ increase. This effect is most noticeable in the *cis* coupling constants between olefinic protons in cycloalkenes **83-86**: as the ring size increases, the coupling constant increases. It is therefore possible to tell in many cases into what size ring, from three- to six-membered, a double bond is incorporated.

$$H_a \quad \theta \; \theta' \quad H_b$$

82

0.5-2.0 Hz   **83**   2.5-4.0 Hz   **84**   5.1-7.0 Hz   **85**   8.8-10.5 Hz   **86**

Bond-length dependence. Double bonds are shorter than single bonds, vicinal overlap is better, and the coupling constants are larger, other things being equal. Thus, cyclohexadiene **87** has similar dihedral angles for all the adjacent olefinic C—H bonds; but the coupling constant is greater across the double bonds than across the intervening single bond. Open chain dienes like butadiene **88** exist mainly in the s-*trans* conformation, and the intermediate coupling constant is greater, but not as large as the *trans* coupling constant for the double bond.

$^3J = 9.4$ Hz
H_a
H_b
$^3J = 5.1$ Hz
H_c

87

H_a H_c
H
$^3J = 17.1$ Hz
H_b
$^3J = 10.4$ Hz

88

H_a
$^3J = 8.5$ Hz
H_b
$^3J = 7.5$ Hz
H_c

89

Aromatic carbon-carbon bonds have bond lengths intermediate between normal single and double bonds. In consequence, *ortho* coupling constants are typically rather lower than *cis* olefinic coupling constants: about 7-8 Hz in benzene rings and 8.8-10.5 Hz in cyclohexenes. We have seen some representative numbers for aromatic compounds in the spectrum of diphenylacetaldehyde **31** in Fig. 3.36 where the *ortho* coupling constant is close to 7 Hz, and in the spectrum of 3-chloropyridine **60** in Fig. 3.34, where the vicinal coupling constants are 8.3 and 4.6 Hz, with the latter reduced in magnitude because of the influence of an electronegative element, the adjacent nitrogen atom. In contrast, alkenes like allyl bromide **37** in Fig. 3.24 and methyl acrylate **42** in Fig. 3.29 have *cis* coupling constants of 10.0 and 10.8 Hz, respectively. Polycyclic aromatic rings have unequal bond lengths, and unequal coupling constants, as in naphthalene **89**.

Tables 3.29 and 3.30 at the end of this chapter summarise vicinal coupling constants.

3.10.2 Geminal coupling ($^2J_{HH}$)

Geminal coupling can only be seen in a spectrum when the two protons attached to the same carbon atom resonate at different frequencies. However, the coupling constants can be measured, even in molecules such as methane, by introducing a deuterium atom, and measuring the geminal coupling from H to D. The value obtained is related to the proton-proton coupling constant by:

$$J_{HH} = 6.5J_{HD} \qquad (3.12)$$

which applies to all coupling, whether geminal or not.

Adjacent π-bonds. The ^2J coupling constant for a simple hydrocarbon, such as methane (measured from a partially deuterated methane), is −12 Hz **90**. When the C—H bonds are able to overlap with neighbouring π-bonds, as in toluene **91** or acetone **92**, orbital overlap is promoted by the adjacent π-system, and the coupling constant is more negative, and effectively larger. The effect is even greater when the hyperconjugation is with the π-bond of a carbonyl group than when it is simply with a C=C double bond—in toluene **91** the coupling constant is −14.3 Hz and in acetone **92** it is −14.9 Hz. The methyl group in toluene and in acetone is freely rotating, and the measured coupling constants are weighted averages of the coupling between the geminal hydrogens for all the conformational relationships in which they find themselves. In rigid, and especially cyclic, systems in which the conformation is held favourably for overlap, with one C—H bond above the π-bond and one below, it commonly reaches −16 or −18 Hz, as it evidently does in the ester **47** with a geminal coupling constant of −17.7 Hz. If the hyperconjugative overlap is with two double bonds flanking a methylene group, the coupling constant can be close to –20 Hz.

Adjacent electronegative elements. In contrast to a π-bond, which is effectively electron-withdrawing, an electronegative element directly attached to the methylene group is effectively a π-donor with respect to the C—H bond, donating electrons into the antibonding C—H σ^*-orbitals **93**. The coupling constant is now more positive, in other words smaller, as we saw in pantolactone **61**, which has a geminal coupling constant of –9.1 Hz.

Note, in contrast, that since the oxygen atom of the carbonyl group of **92** makes the π-orbital of the carbonyl carbon more electron deficient than the corresponding carbon atom in toluene **91**, the geminal coupling is slightly larger in acetone **92**. The general guide that overall electron withdrawal from the bonds connecting coupled protons makes coupling less positive still holds. Because the geminal coupling is negative, the electronegative oxygen increases the magnitude of the geminal coupling in CH_2CO groups, whereas it reduces the magnitude of the positive vicinal coupling in CH—CH—O groups.

Angle strain. An increase in the H—C—H angle makes ^2J more positive, in other words smaller, as we saw in the epoxide **43** with a geminal coupling constant of –2.6 Hz within the ring, in contrast to –12.2 Hz for the methylene group in the side chain. This effect is most noticeable in the methylene groups of terminal alkenes **94**, where the angle is close to 120° and the coupling constant is close to zero, as we saw in methyl acrylate **42** with a geminal coupling constant of –1.2 Hz. This coupling is dependent upon the nature of substituents at the other end of the π-bond, electronegative elements, like fluorine, making them more negative and electropositive elements, like lithium, actually making the coupling positive in sign and quite large. The effect of the H—C—H angle is also seen in the ranges of ^2J for cycloalkanes (Table 3.28).

3.10.3 Long-range coupling ($^4J_{HH}$ and $^5J_{HH}$)

The coupling constants for allylic, W and other long-range coupling were discussed above, where the main influence was the degree of overlap through the intervening bonds. The coupling constants are usually small, and the influence of substituents not all that noticeable. Most visible allylic coupling, ^4J, is negative in sign, but passes zero to low positive as the degree of overlap of the C—H bond with the π-bond changes from maximum (at 0°, as in the drawings **49** and **50**) to minimum (at 90°). Most homoallylic coupling, ^5J, is positive.

3.11 Line broadening and environmental exchange

Even in the absence of spin-spin coupling, the signals in NMR spectra are not lines—they all have appreciable, although often very narrow, half-height widths. There are four common reasons for broadening: unresolved coupling, the more or less inevitable small inhomogeneities in the magnetic field, efficient relaxation and environmental exchange.

3.11.1 Efficient relaxation

When a nucleus loses energy on passing from the high-energy N_β state to the low-energy N_α state, it is said to *relax* (Fig. 3.1). The energy that is lost passes to the surrounding environment, which is called the *lattice*, and the equilibrium Boltzmann population, which existed before excitation, is restored through relaxation. Just as excitation requires the matching of the frequency of irradiation with the natural frequency of the nucleus, so relaxation requires the fluctuation of a local magnetic field with components at the natural frequency of the nucleus. This can be achieved if the molecule containing the nucleus tumbles at a rate that is close to the natural frequency of the nucleus (20-750 MHz, depending upon the nucleus and the field strength). Thus, molecular tumbling rates of ca. 10^8 s^{-1} promote efficient relaxation. The origin of this effect can be seen by considering a model for the relaxation of ^{13}C nuclei by ^{1}H (Fig. 3.43). In Fig. 3.43a, a sample C—H bond is oriented parallel to the applied magnetic field B_0, and the nuclear magnets of the ^{13}C and ^{1}H nuclei are depicted in an orientation that represents their preferred alignment (arrangement of lowest energy) in this field. The lines of force associated with the magnetic field from the nuclear magnet of ^{1}H are indicated by the broken line, which shows that they reinforce the applied magnetic field at the ^{13}C nucleus. If the molecule containing this C—H bond rotates through 90° as the molecule tumbles in solution, the new orientation of the bond is shown in Fig. 3.43b. The notional 'bar magnets' associated with the nuclei are again oriented along the direction of B_0, but now the lines of force from the ^{1}H nucleus oppose the applied field at the ^{13}C nucleus. Thus, as the molecule tumbles in solution, the field at the ^{13}C nucleus fluctuates at the frequency of the rotation. This magnetic field fluctuation provides an efficient mechanism for the relaxation of ^{13}C by ^{1}H when this frequency has components close to the resonance frequency of ^{13}C. In the same way, protons that are close in space to each other in a molecule can relax each other in what is called *dipolar relaxation*.

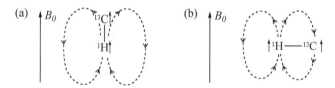

Fig. 3.43

The efficiency of one nucleus to relax another by dipolar relaxation is also dependent upon r^{-6}, where r is the distance between the nuclei, and so a combination of proximate ^{1}H nuclei and relatively slow tumbling promotes relaxation.

The effect of efficient relaxation on the appearance of the NMR spectrum is to broaden the lines. This is a consequence of the Heisenberg uncertainty principle; if a state has a lifetime τ_m, then there is an uncertainty in its energy given by:

$$\delta E = \frac{h}{2\pi\tau_m} \tag{3.13}$$

Since the relaxation rate is τ_m^{-1}, then the larger it is the larger will be δE. However, from Planck's law, $\Delta E = h\nu$ and hence an uncertainty in the energy requires an uncertainty in the frequency. Thus, the faster the relaxation, the broader the line. In small molecules, with molecular weights up to a few hundred, the tumbling frequency is close to 10^{11} s^{-1}— too fast to promote efficient relaxation—and the NMR signals are sharp. However, larger molecules, with molecular weights of, say, 1000 or more, tumble at a frequency of approximately 10^9 s^{-1} or less (the precise value also depending upon the viscosity of the solvent), and the NMR signals are broader. Furthermore, the higher the molecular weight, the broader they are.

3.11.2 Environmental exchange

The signals from protons bonded to oxygen, nitrogen and sulfur are difficult to predict both in chemical shift and in their appearance. We saw in section 3.4.2 that their chemical shift is affected, among other things, by the degree of hydrogen bonding. We have also seen, although it was not commented upon, that OH groups can give rise to very broad signals in a carboxylic acid **32** (Fig. 3.18), and to broad signals in two alcohols, hexanol **38** (Fig. 3.25) and pantolactone **61** (Fig. 3.35). The coupling of the OH proton with the neighbouring hydrogens in hexanol **32** does appear in the methylene signal, which is a quartet, but it does not appear in pantolactone, where H_c is a singlet. The reason for all this variability is that protons attached to any of these three common elements can exchange with the protons of other OH, NH and SH groups, either in other molecules of the same compound or in the solvent, if the solvent has exchangeable protons. The appearance of these signals in the NMR spectrum is affected by the rate of this exchange, which is, in turn, affected by the concentration, the temperature, the nature of the solvent, and by the presence or absence of acid or base catalysis. Coupling to OH, NH and SH protons can appear when the rate of exchange is low, as it is when the sample is in dilute solution, is exceptionally pure, or when the solvent is d_6-DMSO.

In more detail: if the rate of the exchange in an alcohol such as hexanol is appreciably faster than the difference in frequency between the lines of the methylene triplet (6 Hz in this case), the receiver detects the hydroxyl proton in an average of the three magnetic microenvironments created by the three possible arrangements of the nuclear magnets of the methylene protons, and it gives one signal. In very pure samples and in d_6-DMSO, the rate constant for the exchange is less than 6 s^{-1}, and the coupling is then visible; the methylene signal in the spectrum of hexanol in DMSO appears as a quartet, and the OH signal as a triplet; in pantolactone H_c and the OH group give rise to an AX system. The more and more frequent use of very dilute solutions is also making coupling with OH protons increasingly visible. Coupling involving NH protons in amines and SH protons is similarly visible only in special cases. However, CH groups adjacent to amide NH groups usually show coupling (J = 5-9 Hz), even when the amide NH signal is very broad. In such cases, the broadness of the NH signal is not primarily because of exchange-induced broadening, but rather from fast relaxation of the NH proton caused by the quadrupole moment of the ^{14}N nucleus. The signals of the OH, NH and SH groups are removable (Sec. 3.4.2) with a D_2O shake, except for the NH groups of amides, which are slow to exchange in the pH range 2-4.

Similarly, if we have a pair of protons in different environments changing places within the molecule, the signal they give is affected by the rate of interchange. At one extreme, when the rate of exchange is very low, they will appear as separate signals. At the other extreme, when the rate of exchange is very fast, they will appear as a single line. In between, when the rate constant for exchange is comparable to the difference in frequency of the signals, we see broadened lines. A macroscopic analogy is found in a slowly run film allowing us to see an object in two distinct environments, whereas in a quickly run film we may see the object in a time-averaged position. Let us start with two separate signals from protons exchanging between two environments slowly, and let us imagine that the temperature is raised. Initially, we would see the separate signals (Fig. 3.44a) broaden and flatten out (Figs 3.44b and 3.44c), before they coalesce (Fig. 3.44d) and then sharpen again (Figs 3.44e and 3.44f), until they became a single line.

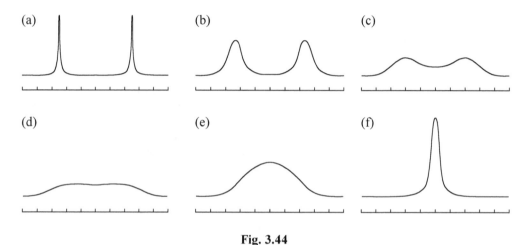

Fig. 3.44

When the two environments are equally populated, and the nuclei are not coupled, the rate constant (s^{-1}) for the exchange at the temperature of coalescence (Fig. 3.44d) is given by:

$$k = \frac{\pi\Delta v}{\sqrt{2}} = 2.221\Delta v \qquad (3.14)$$

where Δv is the difference in frequency between the initially sharp lines. Thus, NMR spectroscopy can be used to measure the rate constants for those events which take place at suitable rates, often loosely referred to as *the NMR time scale*.

For example, the ^{1}H NMR spectrum of dimethylformamide **95**, taken at room temperature, shows two singlets for the *N*-methyl groups at δ 3.0 and 2.84. This is because π-overlap between the nitrogen lone pair and the carbonyl π-bond (**95**, arrows) slows the rotation about this bond. On warming, however, the lines broaden and coalesce, as in Fig. 3.44. The coalescence temperature T_c of 337 K was measured long ago on a

95

60 MHz instrument, from which we can calculate the free energy of activation for the rotation at this temperature using:

$$\Delta G^{\ddagger} = RT_c [23 + \ln \frac{T_c}{\Delta v}] \tag{3.15}$$

where T_c is expressed in Kelvin and R is the gas constant. The answer in this case is 74 kJ mol^{-1}. It is ironic that this particular experiment would be more difficult today, because the coalescence temperature would, inconveniently, be closer to 360 K on a 400 MHz instrument.

3.12 Improving the NMR spectrum

3.12.1 The effect of changing the magnetic field

NMR spectrometers are available with a variety of magnetic fields; thus 200, 250, 300, 400 and 500 MHz instruments are in common use (the resonance frequency for protons is used to identify them), and instruments with ever higher fields are steadily being introduced. These high-field instruments have a number of substantial advantages, which has brought them rapidly into general use, in spite of the cost of superconducting circuitry and the attendant support costs. The frequency of a resonance changes as the field is changed, but the chemical shift value δ does not. This means that the separation between, for example, $\delta = 2$ and $\delta = 3$ is 200 Hz on a 200 MHz instrument, but 400 Hz on a 400 MHz instrument (remember that δ is expressed in p.p.m.). Coupling constants do not change as the magnetic field changes, but their appearance in the spectrum does. Thus, a doublet with a coupling constant of 18 Hz occupies 30% of the space between δ values differing by 1 p.p.m. in the spectrum from a 60 MHz instrument, but only 4.5% of the space on the spectrum from a 400 MHz instrument. This means that multiplets which overlap when the spectrum is taken at low field are much less likely to do so at high field—the multiplet is effectively narrower. Spectra taken on high-field instruments are more likely to be first order, the signals are more easily recognised, and otherwise unresolved signals can come out of methylene envelopes. The effect can be quite dramatic, as seen in the spectra of aspirin **96** taken on two different instruments (Fig. 3.45).

Fig. 3.45b shows an old spectrum of the aromatic region measured at 60 MHz; the four different protons are all coupled to each other, and give rise to a spectrum which, except for H_a, cannot easily be analysed by the first-order approximation. Fig. 3.45a shows the same part of the spectrum measured on a 400 MHz instrument, where the signals are now clearly separate and easily analysed. It is clear why the old instruments operating at 60 MHz have completely given way to the high-field instruments of today.

A second advantage of high-field instruments is their greater sensitivity. At higher field strengths, there is a bigger separation in energy between the two spin states, and a

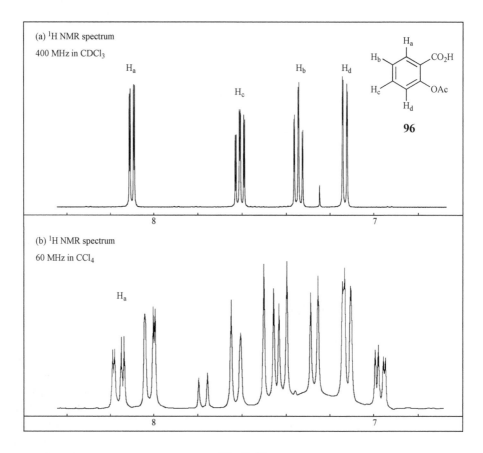

Fig. 3.45

bigger difference in the numbers of the nuclei N_α and N_β. As engineers have been able to design and manufacture instruments operating at higher and higher fields, ever smaller samples can be used, a factor of especial importance in ^{13}C NMR, where the low sensitivity and low natural abundance of the nucleus had limited its use.

A third difference with high-field instruments, not always an advantage, is that the NMR time-scale changes, and the range of dynamic processes that can be studied by NMR spectroscopy is changed.

3.12.2 Shift reagents

In the spectrum of n-hexanol **38** in Fig. 3.25, signals from six of the protons in similar environments overlap in the methylene envelope between δ 1.35-1.18, and their multiplicity can no longer be seen, even in a 400 MHz spectrum. This is a much more serious problem in larger molecules when signals often overlap, causing useful information to be buried. The addition of a shift reagent alters this picture dramatically. Shift reagents are usually β-dicarbonyl complexes of a rare earth metal, the commonest being Eu(dpm)$_3$ **97**, Eu(fod)$_3$ **98** (M = Eu), and Pr(fod)$_3$ **99** (M = Pr). These complexes are mild Lewis acids, which attach themselves to basic sites such as hydroxyl and carbonyl groups. They are also paramagnetic, and have the effect of changing substantially the magnetic field in their immediate environment. The result is a shift of the signals coming from the protons near the basic site in the organic molecule. The shift, downfield with the two europium reagents but upfield with the praseodymium reagent, falls off, with angular variation, as the inverse cube of the distance from the metal. Thus, the spectrum of n-hexanol is spread out when Eu(dpm)$_3$ is added to the solution, and the resonances of each of the three methylene groups can be seen as two quintets and a sextet.

97 Eu(dpm)$_3$ 98 M = Eu or Pr   99

The amount of shift reagent used need not be equimolar, since the Lewis salt is being formed and broken rapidly on the NMR time scale, and a weighted average between the signal of the uncomplexed alcohol and the signal of the Lewis salt is detected. The penalty paid for having a paramagnetic salt present is a broadening of the lines of the multiplets. The multiplicity of the signals is usually clear enough in spectra taken at 100 MHz or less. Unfortunately, the broadening is a function of the square of the operating field strength, and so shift reagents are much less useful on modern instruments, where broadening obscures the multiplicity.

One of the remaining applications of shift reagents is in the measurement of the proportions of the enantiomers present in an incompletely resolved mixture. The traditional method, measurement of the rotation of polarised light, is apt to be misleading if impurities rotate the plane of the polarised light substantially more than the compound under investigation. Furthermore, this method can only be used if the extent of rotation given by one of the pure enantiomers is already known. If the chiral molecule has a basic site, and the shift reagent is an enantiomerically pure complex such as the camphor derivative **99**, the two enantiomers under investigation can have different binding constants and can adopt different conformations on binding, with the result that their NMR signals are shifted to different extents. If the signals separate adequately, they can be integrated, and the proportions of the two enantiomers measured.

3.12.3 Solvent effects

Another kind of shift reagent, although not usually called a shift reagent, is the solvent. By changing the solvent from a solvent of low polarity like deuterochloroform, to a polar solvent like d$_6$-DMSO, or to an aromatic solvent like d$_6$-benzene or d$_5$-pyridine, the chemical shifts of individual signals can change, and overlapping multiplets can be resolved. These solvent-induced shifts, which rely on solute-solvent interactions, though roughly predictable in the case of d$_6$-benzene, are often unpredictable, but they are certainly useful.

Enantiomerically pure chiral solvents are also used to measure the proportions of enantiomers present in an incompletely resolved mixture. Mildly acidic solvents like the fluorinated alcohol **100** can be used to analyse basic compounds like amines, and chiral amines for acidic substances. More commonly, the analysis of mixtures of enantiomers is carried out by attaching a chiral auxiliary covalently to them both, and measuring the

NMR spectrum of the mixtures of diastereoisomers. The most frequently used chiral auxiliary is Mosher's acid **101**, the esters and amides of which give sharp and frequently well-separated signals in the ^{1}H NMR spectrum from the methoxy group, or in the ^{19}F NMR spectrum from the trifluoromethyl group. It is even possible, because the diastereoisomers are so similar, to integrate corresponding pairs of signals in the ^{13}C NMR spectrum.

100   **101**

3.13 Spin decoupling

3.13.1 Simple spin decoupling

In earlier sections we have seen how a proton with neighbouring protons can give rise to a multiplet, how the multiplet pattern can be recognised and how coupling constants can be measured. However, in more complicated molecules than the ones we have seen so far, we want to be able to pair up and extend spin systems that are not obvious. The multiplicity and the chemical shift of a signal from one or more protons may be visible, but it is not always obvious to which of the other signals it is coupled. We were able to pair up the signals from ethyl propionate **33** in Fig. 3.23 using the coupling constants, but we cannot use this technique to pair up the quartets and triplets in diethyl acetylsuccinate **47** in Fig. 3.31, because the coupling constants are the same for both ethoxy groups. Fortunately there is a powerful technique, *spin decoupling*, for making this type of connection unambiguously.

If, during the time that a signal is being collected, the proton (or any other magnetic nucleus) has a neighbour that is exchanging its spin state rapidly, the proton we are observing will experience an average of all the states. We have seen this already in the loss of coupling to the OH proton in pantolactone **61** in Fig. 3.35, where the exchange was a chemical exchange, the OH protons moving from molecule to molecule with a rate constant greater than the coupling constant. The same loss of coupling occurs when the exchange is between spin states stimulated by irradiating the neighbour at its resonance frequency, as we have seen in all proton-decoupled ^{13}C spectra. In those spectra the decoupling was unselective, but it is also possible to decouple selectively.

Thus, we can look again at the two triplets and quartets from diethyl acetylsuccinate **47** (Fig. 3.31 with the undisturbed spectrum repeated in Fig. 3.46). If we measure the resonance frequency ν_1 of the downfield quartet, and then irradiate the sample precisely at that frequency at the same time as we collect the spectrum, the spin states of these methylene protons will rapidly exchange places with each other. The signal at ν_1 appears as a strong distorted singlet with side bands spaced equally above and below this singlet by 0.08 p.p.m., as seen in the expansion in Fig. 3.46. The methyl protons to which they are coupled will have their coupling to the methylene group 'turned off', and they come into resonance at their usual frequency ν_2, but as a singlet instead of a triplet (right-hand enlargement in Fig. 3.46). In this way we can be sure that the methylene group giving rise to the downfield quartet is connected to the methyl group giving rise to the downfield triplet. The irradiation at v_1 leaves other signals more or less undisturbed (such as the signals illustrated in the enlargements in Fig. 3.46 from the other methylene group, from H_a and from the other methyl group), because they are not coupled to the protons being irradiated. There is one limitation: the relevant signals must be reasonably well separated in chemical shift. Thus, in the case of the ester **47**, v_1 is well separated from v_2, and there is no problem there, but the downfield quartet is perilously close in chemical shift to the upfield quartet from the other methylene group. Fortunately, it is just far enough away for the experiment to work, but it would probably not be possible to carry out the experiment the other way round—irradiating one of the triplets to see which quartet collapsed, because the triplets are too close in chemical shift. When the nuclei are the same element, typically both ^{1}H, this technique is called *homonuclear decoupling*. When the experiment involves nuclei of different elements, it is called *heteronuclear decoupling*.

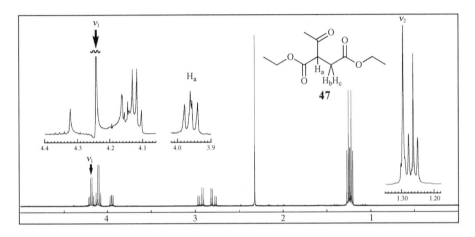

Fig. 3.46

The technique works for all kinds of multiplets, so that a double doublet, for example, will collapse to a doublet when the sample is irradiated at the resonance frequency of one of the protons to which it is coupled. This technique is used a lot when only a few of the coupling relationships are in doubt. In a more complex molecule, it is possible to irradiate successively at the frequency of each of the signals in the spectrum, to plot the spectrum in each case, and to look at all the spectra to find which signals lose coupling in each experiment. This can reveal all the coupling relationships, and all the connections between the various protons within a molecule. It is now more usual, however, if a lot of coupling information is needed, simply to run a COSY spectrum to achieve this end (see Sec. 3.19).

3.13.2 Difference decoupling

It is possible to use selective decoupling to reveal a buried signal. Fig. 3.47c shows a narrow part of the methylene region, between δ 1.14 and 0.9, of the spectrum of the steroid **102**. This signal is a composite of the multiplets from four protons, one of which is $H_{7\alpha}$, all overlapping inextricably. When the signal of $H_{6\alpha}$, which is further downfield than the ones in Fig. 3.47, is irradiated, the signal from $H_{7\alpha}$ loses one of its couplings, and the signal changes to that in Fig. 3.47b. The multiplets are just as impossible to analyse as before, but now, because the spectrum is in the computer in digital form, it is possible to subtract the original spectrum (c) from the decoupled spectrum (b). The result is plotted in Fig. 3.47a. Fortunately, the other three protons in this signal were not coupled to $H_{6\alpha}$, and were unaffected by the decoupling; the subtraction therefore removed them from the signal and left only a signal from $H_{7\alpha}$. The signal left has both the coupled and the decoupled signal in it, the original coupled signal is a double quartet down and the partly decoupled signal is a quartet up, since (c) was subtracted from (b). Evidently $H_{7\alpha}$ is coupled equally to each of the protons $H_{6\beta}$, H_8, and $H_{7\beta}$, with a coupling constant of 13 Hz, leading to the quartet, and to $H_{6\alpha}$ with a coupling constant of 4.3 Hz, doubling that quartet.

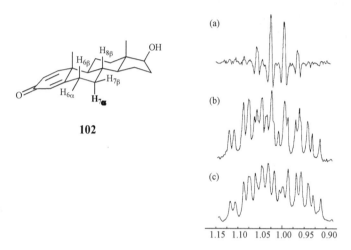

Fig. 3.47

(Reproduced with permission from J. K. M. Sanders and B. K. Hunter, *Modern NMR Spectroscopy*, OUP, Oxford, 1987.)

3.14 The nuclear Overhauser effect

3.14.1 Origins

The interaction of one magnetic nucleus with another leading to spin-spin coupling takes place through the bonds of the molecule. The information is relayed by electronic interactions, as one can see from the dependence of the coupling constant on the geometrical arrangement of the intervening bonds.

Magnetic nuclei can also interact through space, but the interaction does not lead to coupling. The interaction is revealed when one of the nuclei is irradiated at its resonance frequency and the other is detected as a more intense or weaker signal than usual. This is called the nuclear Overhauser effect (NOE or nOe). The NOE is only noticeable over short distances, generally 2-4 Å, falling off rapidly as the inverse sixth power of the distance apart of the nuclei. This is because the interaction is dependent upon the relaxation of the observed nucleus by the irradiated nucleus (Sec. 3.10).

Two nuclei A and X relaxing each other, but not coupling, interact to set up four populated energy levels, as we saw in Fig. 3.38b. We shall ignore the complication of coupling here, because it is quite separate from the NOE, and does not interfere. The NOE is through space, and nuclei can show NOEs whether they are coupled or not.

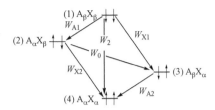

Fig. 3.48

Fig. 3.48 repeats Fig. 3.38b and augments it. The pairs of arrows depict the orientations of the nuclear magnets of the A and X nuclei in the applied magnetic field. The transitions W_{A1} and W_{A2} lead to the line we associate with the A nucleus and the transitions W_{X1} and W_{X2} lead to the line we associate with the X nucleus. If the sample is irradiated at the resonance frequency of the X nucleus, the transitions W_{X1} and W_{X2} take place rapidly in both directions. The population levels (1) and (2) grow at the expense of levels (3) and (4), respectively, so that levels (1) and (3) are equally populated, and so are levels (2) and (4). There is still no obvious effect on the intensity of the A signal, because it is produced by transitions from (1) to (2) and from (3) to (4), and the former has increased at the expense of the latter. The intensity of the A signal is dependent upon the difference between the sum of the populations of (1) and (3) and of (2) and (4), and this has not been affected. However, there are two other relaxation pathways, W_2 and W_0, which do not lead to observable signals but do affect the populations of the four energy levels. W_2 is a two-quantum process (two nuclei change their spins) between well-separated energy levels, and relaxation by this pathway is stimulated by the more rapid (higher-frequency) tumbling of molecules with a molecular weight in the region of roughly 100-400. The effect is to increase the population of energy level (4) at the expense of energy level (1). The sum of the populations of the energy levels (1) and (3) is reduced relative to the sum of the populations of the energy levels (2) and (4), and the signal from the A nucleus is, therefore, increased in intensity. When the W_2 relaxation pathway is dominant over the W_0 pathway, the intensity of the observed resonance of A is therefore increased, and this effect is called a *positive NOE*.

In contrast, W_0 is a zero-quantum process (no net change of spin) between energy levels close in energy, and relaxation by this pathway is stimulated by the slower (lower-frequency) tumbling of larger molecules with molecular weights ≥ 1000. The effect is to increase the population of energy level (3) at the expense of energy level (2). The sum of the populations in energy levels (1) and (3) is now increased relative to the populations of levels (2) and (4), and the signal from the A nucleus is reduced in intensity—a *negative NOE*. Molecules with intermediate molecular weight fall between two stools, and show weak or non-existent NOEs.

In summary, small molecules with molecular weights up to about 300, in solvents of the viscosity most commonly used in NMR, tumble with a frequency close to 10^{10}-10^{11} s^{-1} and the NOE is an enhancement. In large molecules, with molecular weights greater than 1000 (or even greater than around 600 in a relatively viscous solvent such as DMSO), the NOE corresponds to a diminution of the observed signal. The former is referred to as a *positive* NOE and the latter as a *negative* NOE.

In ^{13}C spectra, the maximum possible NOE produced by irradiating at the proton frequency is nearly 200%. NOEs of this order are found in the signals from ^{13}C atoms directly bonded to protons in proton-decoupled spectra, and are helpful in increasing the intensity of otherwise inherently weak signals. In ^{1}H spectra, the maximum enhancement can only be 50% of the usual intensity of the signal, but the typically observed range is only 1-20%. NOEs are weakened when the proton being observed is being relaxed by protons other than the irradiated proton. Thus, a methyl group, in which each proton already has two nearby protons to speed up relaxation, often shows very little NOE when a nearby proton is irradiated. NOEs are most easily detected, therefore, in methine groups. It is possible to measure NOEs by integrating signals with the irradiation on and then with it off, and measuring the difference by integration. The accuracy of integration is such that this method can only be used reliably if the NOE is at least 10%, and it can only be detected in the signals from methine and, occasionally, methylene protons. Nevertheless, even in this form, the NOE has been a useful method for detecting which groups are close in space to each other, providing valuable information about stereochemistry, but it has been superseded by difference and NOESY spectra (respectively, below and Sec. 3.19).

3.14.2 NOE Difference spectra

NOEs are much more easily detected by subtracting, in the computer, the normal spectrum from a spectrum taken with the irradiating signal on, and printing only the difference between the two spectra. All the unaffected signals simply disappear, and all that shows is the enhancement itself, together with an intense signal at the irradiating frequency. The lower trace in Fig. 3.49 shows the complex ^{1}H spectrum of the oxindole **103**, and the upper trace is the difference spectrum created after irradiating the sample at the frequency of the heavy downward-pointing arrow. This frequency is that of proton H_{7a}, which is close in space both to its neighbour H_{7b} and to $H_{5'}$ on the benzene ring. Only signals from these two protons appear in the difference spectrum, and demonstrate that the stereochemistry of the molecule **103** is that shown, and not the isomer with the spiro-oxindole ring the other way up. When a similar experiment is carried out on that isomer, no signal appears in the aromatic region of the difference spectrum. The signal from H_{7b}

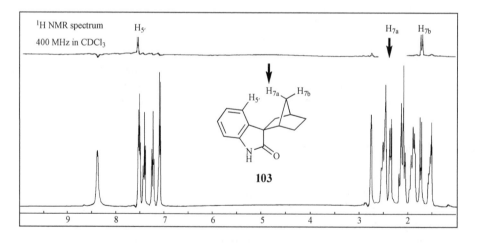

Fig. 3.49

in the difference spectrum (Fig. 3.49) still shows the coupling to H_{7a}; this is because the signal used to create the NOE is applied before the acquisition pulse, but is turned off during acquisition. Coupling is therefore unaffected.

Using difference spectra, it is easy to detect 1% enhancements, or even less, with the result that NOEs in methyl groups are now quite commonly measurable. In consequence, it is usually possible to detect the NOE in both directions—not only, for example, from a methyl group to a nearby methine, but also back from the methine to the methyl group, a procedure that greatly increases one's confidence that the groups are indeed close to each other. Furthermore, the distance over which the NOE can now be detected is much greater, and more structural information is available for that reason too.

Difference NOE spectra also allow one to extract a signal from under several others, in exactly the same way as for decoupling difference spectra. The lower spectrum in Fig. 3.50 is the methylene region of the 6-methylprogesterone **104**, and above it is the difference spectrum produced after irradiating at the resonance frequency of the C-19 methyl group.

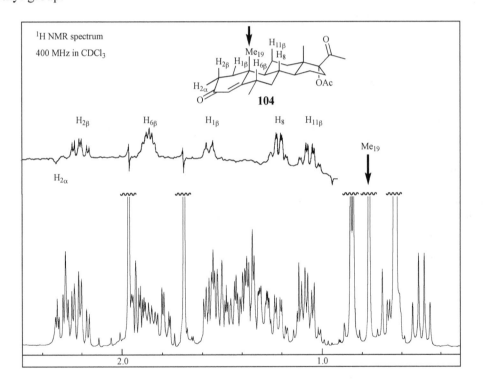

Fig. 3.50

(Reproduced with permission from J. K. M. Sanders and B. K. Hunter, *Modern NMR Spectroscopy*, OUP, Oxford, 1987.)

The signals in the difference spectrum come from the group of protons on the top of the molecule surrounding the 19-methyl group, each showing the multiplicity appropriate to it. In the conventional spectrum, only $H_{2\beta}$ is clearly resolved, and even that signal is more securely assigned than it was before, now that it is shown to come from a proton close in space to the C-19 methyl group. The signal from $H_{2\alpha}$ is also evident as a negative NOE.

This is a common observation when a proton, $H_{2\beta}$ in this case, shows a positive NOE, and is relaying an NOE to other protons near it.

3.15 Assignment of CH$_3$, CH$_2$, CH and quaternary carbons in ^{13}C NMR

The pulse and acquisition technique used in FT spectroscopy makes it possible to carry out much subtler experiments than the simple one described in Sec. 3.1. It is possible to give a second or third pulse on the same or a different axis, and it is possible to wait for various lengths of time between pulses; the pulses can be $\pi/4$ pulses, or π pulses, or any other angle, as well as the $\pi/2$ pulse illustrated in Fig. 3.2. It is not possible to describe in the following sections of this chapter the pulse sequences, nor what their effect is on the precessing nuclear magnets; but several books listed in the bibliography at the end of this chapter do that. We shall simply take the output of these experiments, illustrate them, and give a brief account of those you are most likely to encounter, so that you can recognise them, and know what kind of information can be obtained from them.

In structure determination, it is extraordinarily helpful to identify the number of CH$_3$, CH$_2$, CH and C atoms with no attached hydrogen atoms (often loosely called quaternary carbons, although this term ought to be restricted to C atoms with four carbon substituents). Although this can be done by off-resonance decoupling in a ^{13}C spectrum (Sec. 3.5), it is more effectively achieved using a multi-pulse sequence. The most straightforward multi-pulse sequence gives an APT (Attached Proton Test) spectrum, in which the CH$_3$ and CH carbons give signals in one direction and the CH$_2$ and fully substituted carbons give signals in the other. The APT spectrum of 3,5-dimethylbenzyl methoxyacetate 1 is shown in Fig. 3.51, which can be compared with the normal proton-decoupled spectrum in Fig. 3.5. In the APT spectrum, the carbon atoms having an even number of attached hydrogen atoms (0 and 2) produce positive signals, and those with an odd number of hydrogens (1 and 3) produce negative signals. APT spectra are not always plotted this way up—there is no rigid convention, but which way up they are plotted can be deduced from the signal of deuterochloroform, which has no protons and so gives

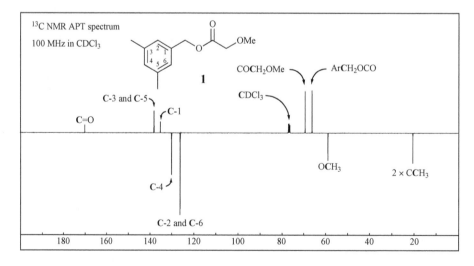

Fig. 3.51

signals in the same direction as the other fully substituted carbons and methylenes. Although APT spectra do not distinguish between methyl and methine, nor between methylene and fully substituted carbons, it is rare for there to be much ambiguity. Methyl and methine carbons can usually be distinguished by their chemical shift, and methylene and fully substituted carbons by their intensities. The assignment of the digonal carbons of acetylenes (as CH *vs.* C) from APT spectra is unreliable. This lack of reliability arises because the 'up' or 'down' nature of the observed signals depends upon the magnitude of the C—H coupling constant (where present). In the APT pulse sequence, if a delay is set to optimise the reliability of the data for tetrahedral and trigonal C—H coupling, then the data for acetylenic CH *vs.* C may be misleading.

However, ambiguities do occur from time to time in APT spectra. A multi-pulse sequence which leaves no ambiguity is called DEPT (Distortionless Enhancement through Polarisation Transfer). In this sequence, three spectra are taken in which pulses at three different angles are applied to the protons, influencing the way the protons transfer polarisation to the carbons. The angles are 45°, 90° and 135°, giving the three spectra in Fig. 3.52, from the same ester **1**. The DEPT-45 spectrum 3.52a shows positive CH_3, CH_2, and CH signals; in other words everything except the fully substituted carbons: the C=O carbon, C-1, C-3 and C-5. The DEPT-90 spectrum 3.52b shows only the CH signals, or rather shows only those signals strongly, since there are small signals from other carbons, stemming from a not uncommon but imperfect choice of parameters in the pulse sequence. And the DEPT-135 spectrum 3.52c shows the CH and CH_3 signals positive, but

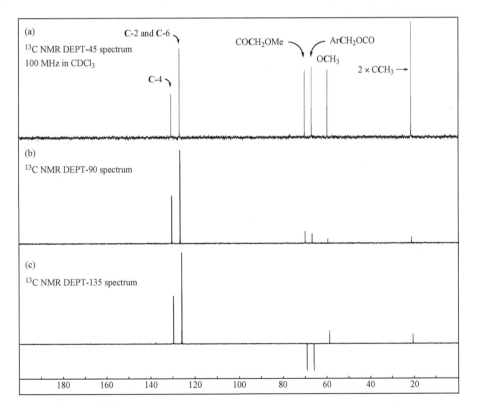

Fig. 3.52

the CH_2 signals negative. The spectrum 3.52b allows us to identify unambiguously the signals derived from CH groups, and the spectrum 3.52c allows us to identify unambiguously the signals derived from CH_2 groups. Both spectra 3.52b and 3.52c reveal that those signals still left unaccounted for in Fig. 3.52a are derived from the CH_3 groups, and those signals still left unaccounted for in the full spectrum (Fig. 3.5), not appearing in any of the three spectra in Fig. 3.52, are derived from the fully substituted carbons. It is not necessary to process these spectra further, since we have already learned all that we want to know. However, it is possible, by weighted additions and subtractions of one spectrum from another to generate three spectra showing a separate ^{13}C spectrum for each of the CH_3, CH_2 and CH signals, and DEPT spectra are sometimes shown in this form.

3.16 Identifying spin systems—1D-TOCSY

In the ketone **105** the protons on C-1, C-2 and C-6 are coupled to each other, and so are the protons on C-4, C-5 and C-7. For both of these systems, summarised in the part structure **106**, each of the diastereotopic protons H_A and H_B on the methylene groups will be double doublets, the methyl groups will be doublets, and the methine signals will be double-double quartets. However, the carbonyl group in between insulates the two sets of signals from each other, and there is no coupling from the one system to the other. The *spin systems* in this ketone are the two $AA'MX_3$ systems **106**, an $AA'BB'$ aromatic system and an AB system for the benzyl protons on C-8, each of which is effectively 'insulated' from the others. The full spectrum is shown in Fig. 3.53, where most of the signals are clearly resolved, but only the aromatic protons on C-10 and C-11, the diastereotopic protons on C-8, and the methoxy signal can be easily assigned using their chemical shifts. Additionally, the one proton resonance at δ 2.45 can fairly confidently be assigned to the OH proton on the basis of its broadness.

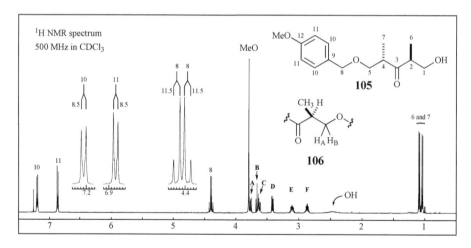

Fig. 3.53

We are left with the two doublets from the methyl groups, C-6 and C-7, two complex signals labelled **E** and **F** from the two methines, and four double doublets, labelled **A-D** from the methylene protons, but we do not know at this stage which signals make up the two $AA'MX_3$ systems, nor do we know which is which. We can try looking at the 'coupling constants', which are summarised on the expanded signals in Fig. 3.54. As we do so, we must bear in mind that these are here measured directly from the spectrum, and therefore may not be fully self-consistent. There are two potential problems. First, there may be a small error in measuring the values; second, since the chemical differences among the coupled protons are comparable to the coupling constants between them (where both are measured in Hz), observed splittings may differ from the true coupling constants.

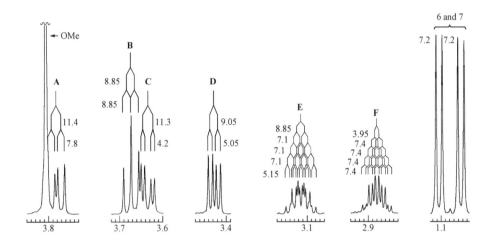

Fig. 3.54

Indeed, the numbers, although close, are not a perfect match. For example, starting with the signal **A**, the measured splittings are 11.4 and 7.8 Hz. These are closest to the larger splitting of 11.3 Hz in **C**, and the larger splitting of 7.4 Hz in **F**, respectively, than to anything else, but they are not identical. We can see the problem again in the methine signal **F**, which approximates to a double quintet. The splitting of 3.95 Hz is probably the counterpart to the splitting of 4.2 Hz in **C**, and the remaining splittings of 7.4 Hz must be to one of the methyl groups (both of which have splittings measured as 7.2 Hz from direct measurements on the methyl group signals). Thus, it is highly probable that we have solved part of the problem: the signals **A**, **C** and **F** make up one of the AA'M systems. Similarly, the rough matching of splittings can identify the **B**, **D** and **E** signals as the other AA'M system, with the double doublet of the **B** signal actually appearing as a triplet, because its two splittings are very similar.

Although all the **A**-**F** signals have identifiable multiplicities, and are consistent with the chemical structure, the assignments are still incomplete. In order to complete the assignments, we can apply a pulse sequence which gives a 1D-TOCSY spectrum (TOtal Correlation SpectroscopY), also called a HOHAHA spectrum (HOmonuclear HArtmann-HAhn). This pulse sequence is designed separately to identify each spin system in a molecule. It requires that a signal from one of the spin systems is separated far enough from the others for it to be selectively irradiated. This is the case for both methine signals from the ketone **105**, one from each of the spin systems. The pulse sequence begins with a long irradiation of the **E** signal at δ 3.12, followed by a mixing phase which causes any protons which are coupled to it to be raised to the N_β level. Let us call these directly coupled protons 'set T'. Depending upon how long the mixing phase is maintained, any protons coupled to 'set T' are the next similarly raised. The FID is then collected in the acquisition phase, and processed to give the spectrum in Fig. 3.55b, showing only the mutually coupled set of protons connected to the **E** signal as well as **E** itself. Similarly, the pulse sequence using selective irradiation at the **F** signal at δ 2.88 brings up the spectrum in Fig. 3.55c, showing only the mutually coupled set of protons connected to the **F** signal as well as **F** itself. We can now see that the signals **B**, **D**, **E** and the upfield methyl doublet make up one of the AA'MX$_3$ systems, and the **A**, **C**, **F** and the downfield methyl doublet make up the other.

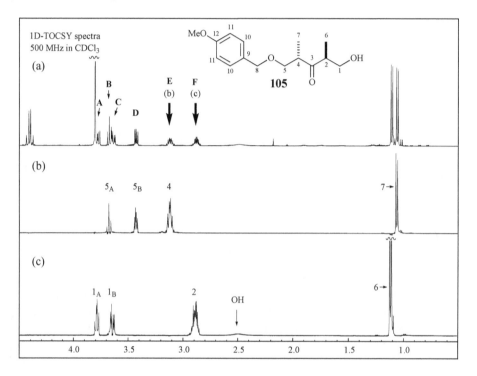

Fig. 3.55

In this particular case, the OH signal appears in Fig. 3.55c but not in Fig. 3.55b, showing that the **A**, **C**, **F** and downfield methyl doublet are produced by the protons on C-1, C-2 and C-6, and therefore the **B**, **D**, **E** and upfield methyl doublet are produced by the protons on C-4, C-5 and C-7. Had we had a molecule without a proton on the C-1 oxygen atom, we would still be unable to identify which side of the molecule belonged to which set of signals. To find out that information we would need the HMBC spectrum, which is described later in this chapter, and which confirms the assignment in Fig. 3.55.

Not only has the 1D-TOCSY spectrum separated out these spin systems, but it has also effectively separated them from signals given by impurities. 1D-TOCSY spectra can therefore be used on mixtures, to pull out at least part of the spectrum of one component. Similarly, signals that overlap the ones in which we are interested, but are not from protons coupled to that spin system, disappear. Thus, the methoxy signal is no longer present in Fig. 3.55c, thereby revealing the whole of the double doublet **A** from H-1$_A$, which is partly obscured by the methoxy signal in the full spectrum.

3.17 The separation of chemical shift and coupling onto different axes

The conventional NMR spectrum is called a one-dimensional spectrum because it has only one frequency dimension—the chemical shift and the coupling are displayed on the same axis with intensity plotted in the second dimension. With larger molecules, this can lead to very complicated spectra, with many overlapping multiplets, even at high field, and with little hope that a shift reagent will help much. For example Fig. 3.50, which we saw earlier, showed the high-field region of the conventional 400 MHz spectrum of 6-methylprogesterone **104**, which has 24 different kinds of proton, most of them in methylene groups. Using a particular pulse sequence, it is possible to collect information and to display it in two dimensions, so that chemical shift is on the conventional axis and the coupling information is on a new axis stretching behind the conventional spectrum. This is called a 2D *J*-Resolved spectrum. A front view of such a plot for the steroid **104** is shown in Fig. 3.56, which now has 22 signals (and some noise), one for each of the different protons coming into resonance between δ 2.5 and 0.5 (H_4 and $H_{16\beta}$ are at lower field). It is a ^{1}H NMR spectrum with all the proton-proton coupling removed, except that the signals are of different intensity from each other because they are projections only of the tallest of the components of each multiplet.

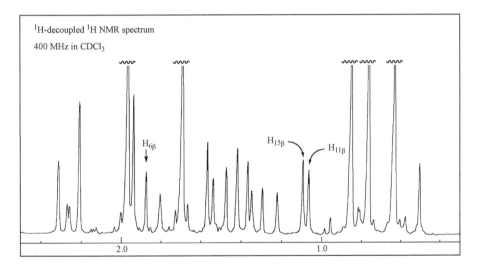

Fig. 3.56

(Reproduced with permission from J. K. M. Sanders and B. K. Hunter, *Modern NMR Spectroscopy*, OUP, Oxford, 1987.)

The multiplicity associated with each of these 'singlets' is contained in the orthogonal 'J-axis', which 'goes down' into the plain of the paper. If these signals are separately plotted for each of the 24 unique protons in the steroid, then the multiplicity of each of these protons is revealed. Three examples, picked out from Fig. 3.56, are illustrated in Fig. 3.57, to show how signals that are hopelessly overlapped in Fig. 3.50 can be extricated using this technique, and their multiplicity revealed. This is our first example of *two-dimensional NMR* spectroscopy—with the chemical shift in one dimension and coupling in the other (two-dimensional because we don't count the intensity, which is always in the third dimension). There are many other pulse sequences giving two-dimensional NMR spectra, greatly extending the range of information that can be pulled out of an NMR spectrum. The rest of this chapter will be concerned with the most useful 2D experiments. 2D-Spectra often need long acquisition times and larger amounts of material (Table 3.46), and are therefore taken only when they are needed—which is often the case for large molecules that give rise to many signals.

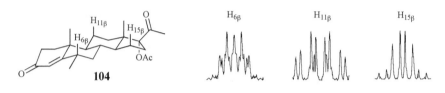

Fig. 3.57

(Reproduced with permission from J.K.M. Sanders and B.K. Hunter, *Modern NMR Spectroscopy*, OUP, Oxford, 1987.)

3.18 Two-dimensional NMR

We have seen in Sec. 3.13 that a full set of difference decoupling experiments would give information about CH_3, CH_2 and CH groups which are coupled to each other. Similarly in Sec. 3.14, we have seen that a full set of difference NOE experiments would define some of the spatial relationships within a molecule. These 1D experiments are still frequently carried out, because they are quick, they are often perfectly adequate in giving the information needed, and they do not need much material. Nevertheless, newer techniques have been invented to determine all the coupling information in a single experiment, or all the spatial information in a single experiment. Both of these powerful experiments are further examples of *two-dimensional NMR* spectroscopy.

In Sec. 3.17 we introduced the concept of plotting the coupling constant on an axis orthogonal to the chemical shift, thus generating a two-dimensional spectrum (ignoring the dimension of intensity). In an analogous way, two-dimensional spectra can be generated where the chemical shift is plotted on two orthogonal axes. When these spectra have been obtained, using multi-pulse sequences, they can be displayed in a manner which bears analogy to a contour map. The two sides of the 'map' are the two orthogonal chemical shift axes, with the intensities of the peaks indicated by the contours, in the same way that the heights of mountains on maps are displayed. There are phase problems to be dealt with, but usually the contour plot uses the absolute values of the intensity of the signals, ignoring their sign in the output from the multi-pulse sequence.

In all of the two-dimensional spectra that we shall describe, multi-pulse sequences are applied such that in the first part of the experiment the magnetisation is changed from its equilibrium state in a specific manner ('prepared'). It is then allowed to evolve as a function of time ('evolution'), and the spins allowed to affect each other's behaviour ('mixing', e.g. according to whether they are spin-spin coupled or relax each other through space). In the last part of the multi-pulse experiment, the resulting magnetisation is detected by collecting FIDs ('acquired'). The experiments can be arranged so that in one kind of pulse sequence (COSY) magnetisation is transferred between nuclei that are spin-spin coupled, or in another (NOESY) magnetisation is transferred between nuclei that undergo mutual dipolar relaxation.

3.19 COSY spectra

A two-dimensional experiment which indicates all the spin-spin coupled protons in one spectrum is called a COSY spectrum (COrrelated SpectroscopY). The COSY spectrum of the α,β-unsaturated ester **107** is reproduced in Fig. 3.60 on the next page, but before we look at that spectrum we must first look at the 1D spectrum in Figs. 3.58 and 3.59, in order to identify as far as we can which signals are produced by which protons.

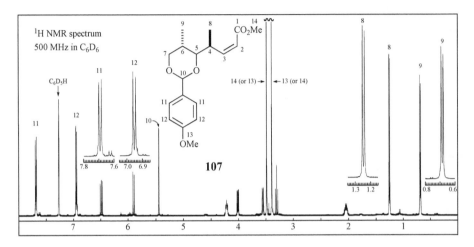

Fig. 3.58

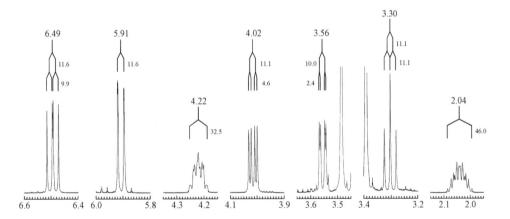

Fig. 3.59

There is little ambiguity about the signals shown expanded in Fig. 3.58, although we cannot be quite sure that the assignment to the protons on C-8 and C-9 is the right way round. The remaining multiplets are shown expanded in Fig. 3.59, where we can be confident, from the coupling and chemical shifts, that the proton on C-3 gives rise to the downfield double doublet at δ 6.49, and the proton on C-2 gives rise to the upfield doublet at δ 5.91. The remaining signals however are not unambiguously assignable. The methine multiplets at δ 4.22 and 2.04 resist a first-order analysis, and so we cannot easily identify their components, nor identify with certainty to which protons they are coupled. However, their very complexity suggests splitting through a large number of interactions, and we can tentatively conclude that the allylic methine proton on C-4 gives rise to the downfield multiplet at δ 4.22, and the methine proton on C-6 gives rise to the upfield multiplet at δ 2.04.

The three one-proton resonances between δ 3.2 and 4.1 must correspond to the non-equivalent methylene pair attached to C-7 and the methine proton attached to C-5. For a complete assignment, we turn to the COSY spectrum in Fig. 3.60.

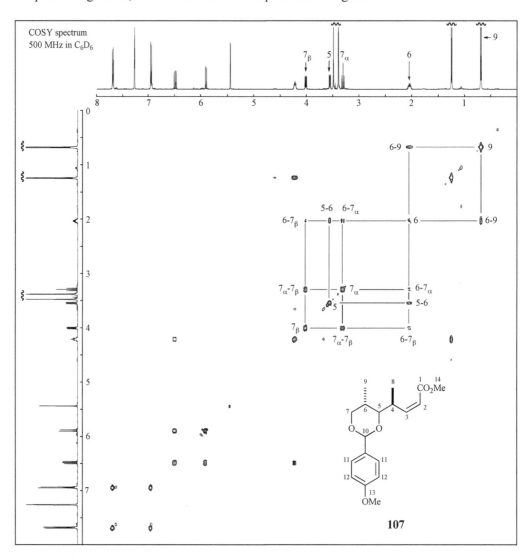

Fig. 3.60

In a COSY spectrum, two identical chemical shift axes are plotted orthogonally, and the one-dimensional spectrum, viewed from above, appears on the diagonal from the bottom left to the top right of the resulting square. For convenience in the interpretation, a 1D spectrum is printed on each of the orthogonal chemical shift axes, and aligned with the spectrum on the diagonal. All peaks that are mutually spin-spin coupled are shown by cross-peaks, which are symmetrically placed about the diagonal. Thus, the cross-peaks labelled 7_α-7_β, making up a square, identify the geminal coupling between the 7_α proton at δ 3.30 and the 7_β proton at δ 4.02, and the cross-peaks labelled 6-7_α and 6-7_β identify the coupling between the methine proton on C-6 at δ 2.04 and the 7_α and 7_β protons at δ 3.30 and 4.02. Likewise, the cross-peaks labelled 6-9 identify the coupling between the methine proton on C-6 and the methyl protons on C-9 at δ 0.68. In this way we have confirmed our guesses that the upfield multiplet at δ 2.04 is produced by the proton on C-6, and that the upfield doublet at δ 0.68 is produced by the methyl protons on C-9. We can move on, using the cross-peaks labelled 5-6, to confirm that the remaining double doublet at δ 3.56 is from the proton on C-5, which is coupled to the proton on C-6. The couplings we have identified so far are summarised in Fig. 3.61a, which shows that the large splittings of 11.1 Hz from the 7_α proton and 10 Hz from the proton on C-5 to the proton on C-6 are appropriate for axial-axial couplings in a chair conformation. The sum of all the splittings (approximate coupling constants) shown on Fig. 3.59 for the multiplet at δ 2.04 is 46.1 Hz, which is, as it should be, close to the total width of this signal shown in Fig. 3.59.

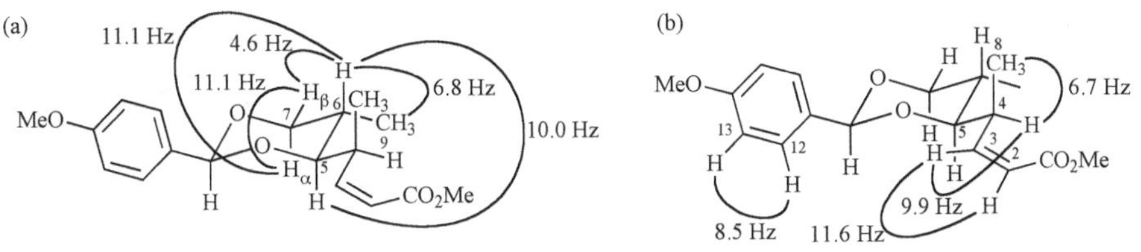

Fig. 3.61

In order to keep the picture simple, the other cross-peaks in Fig. 3.60 are not labelled, nor are the square boxes drawn. We can, however, see that there are cross-peaks for the mutual coupling between the aromatic protons on C-11 and C-12, and similarly for the alkene protons on C-2 and C-3, with the downfield double doublet from the latter having further cross-peaks to the multiplet at δ 4.22 given by the proton on C-4. The signal from the proton on C-4, in turn, has cross-peaks to the methyl protons on C-8 at δ 1.25, completing the assignment. These couplings are summarised in Fig. 3.61b. However, there is one pair of cross-peaks missing—that between the signals at δ 4.22 and 3.56, which ought to show coupling, since they come from the protons on adjacent carbons C-4 and C-5. (The small peaks near to where the cross-peaks ought to be are not quite in the right places—they must have come from a strong coupling in an impurity, the weak peaks from which can be seen in the diagonal in Fig. 3.60, and in the spectra in Figs. 3.58 and 3.59.)

The absence of cross-peaks cannot be taken to mean there is no coupling, only that the coupling constant is small. Thus, the 4-5 cross-peak is missing because of the small coupling constant of 2.4 Hz from the proton on C-5 at δ 3.56, which we have not paired up with anything yet. It must be the remaining coupling to the multiplet at δ 4.22 from the proton on C-4, and if we add up the splittings shown on Fig. 3.59 for all the coupling we have deduced for the multiplet at δ 4.22, and add in the 2.4 Hz, it comes to 32.4 Hz, which is close to the total width of this signal shown in Fig. 3.59. The other detectable coupling with a low coupling constant is the allylic coupling between the protons on C-2 and C-4. It is almost resolved in the signal at δ 5.91 for the C-2 proton, but there is no cross-peak for this either. Long-range coupling can be revealed in a COSY spectrum quite easily, by choosing a lower plane with which to cut the peaks giving the contours, or by changing the pulse sequence to include a delay, which needs to be of the order of $^1/_2 J$. The resulting spectra, necessarily more complicated and noisy, show the long-range coupling. Whatever one is investigating, the planes used to define the contours have to be chosen carefully if all the cross-peaks are to be displayed without too much noise.

When two (or more) signals have very similar chemical shifts, irradiation of one of the signals in the 1D experiment inevitably hits the other at the same time, and we are then unable to tell whether or not they are coupled to each other. COSY spectra are not as limited in this way: as long as the signals are resolved, the cross-peaks can be seen, but there is a problem when peaks on the diagonal are so strong that they obscure the cross-peaks close to the diagonal. There are refinements in the pulse sequence which effectively lower the intensity of the diagonal, and allow this coupling to be identified.

The cross-peaks in COSY spectra contain the coupling constants, but not the full multiplicity. Whereas the signal from the 7_α proton at δ 3.30 is a triplet on the diagonal, the cross-peaks do not have the centre line of the triplet in either direction. This is because the two peaks that do show up in the cross-peak are actually one negative and one positive, with a node in the middle. The same loss of the central line occurs with quintets, but doublets and quartets remain as doublets and quartets in the cross-peaks. The structure of cross-peaks is rarely examined in routine applications of COSY spectra, but there is useful information there for the expert. The COSY experiment, even at the level presented here, is so powerful that it is the most often used 2D technique, routine in many laboratories today.

3.20 NOESY spectra

A two-dimensional spectrum which records all the proton-proton NOEs occurring in a molecule in a single experiment is called a NOESY spectrum. Superficially, it looks like a COSY spectrum—the normal spectrum appears on the diagonal and both orthogonal axes usually have the ^{1}H NMR spectrum printed on them. However, it differs crucially in that the cross-peaks now indicate those protons that are close in space, just as 1D NOE difference experiments detect through-space interactions. Thus, a NOESY spectrum provides information about the geometry of molecules.

The positive NOE enhancements from small molecules (ca. 100-400 Daltons) are rather weak, and these are often more conveniently examined by NOE difference spectroscopy (Sec. 3.13). However, as an illustration of a NOESY spectrum in a relatively simple molecule, let us use the same α,β-unsaturated ester **107** that we looked at in the COSY spectrum. The NOESY spectrum is shown in Fig. 3.62, which you can immediately see, contains a few signals from 'noise'. The correspondence of cross-peaks to noise is indicated, or should be suspected, where (i) there is no corresponding peak in the 1D spectrum, and/or (ii) the cross-peak is not accompanied by a partner which is its reflection point about the diagonal running from bottom left to top right. For example, the diagonal signals at 4.4 and 4.7, and the 'cross-peaks' which give the impression of 'vertical lines' at the chemical shifts (horizontal scale) of the intense methyl signals appearing at δ 1.25, 3.38 and 3.5, should be ignored in the interpretation. We shall see that the NOESY spectrum allows us to confirm all the stereochemistry, especially that at C-10, which was not established by the COSY spectrum.

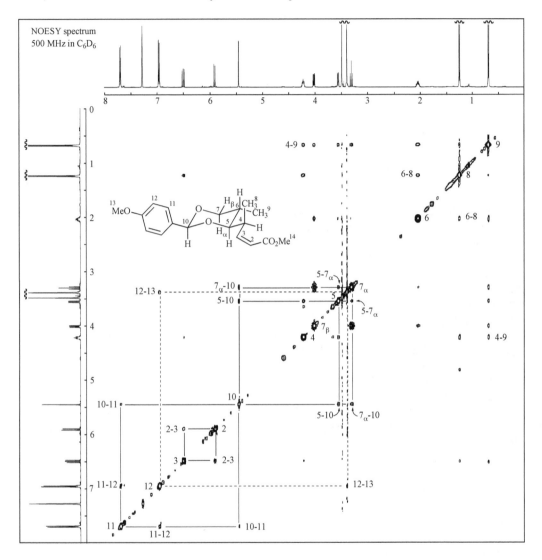

Fig. 3.62

All protons showing geminal, and most showing vicinal coupling, also spend enough time close enough in space to give cross-peaks to each other in the NOESY spectrum. It is the *additional* cross-peaks, labelled in Fig. 3.62, that give us new information. Thus, in this case, the cross-peaks labelled 5-10, 7_α-10 and 5-7_α show that these three protons are close in space because they are all axial. The 4-9 and 6-8 cross-peaks are strong, the 3-4 cross-peaks are weak, but there are no 4-6 cross-peaks, indicating that the molecule spends most of its time in the conformation drawn in Fig. 3.63a, which illustrates the NOEs discussed so far.

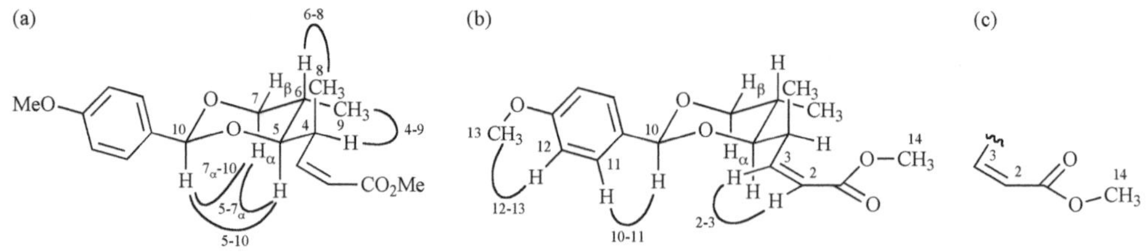

Fig. 3.63

The other informative NOEs are shown in Fig. 3.63b. We can assign the methoxy peaks at δ 3.49 and 3.39 to the ester and ether groups, respectively, because of the presence of the cross-peaks 12-13 connected by the dashed lines. The ester methyl group will be oriented away from the rest of the molecule in the lower energy s-*trans* conformation about the carbonyl-oxygen single bond, as shown in Fig. 3.63b and 3.63c, and does not show an NOE. There is little doubt about the assignment of the aromatic protons, since their chemical shift values are definitive, but the assignment is supported by the 10-11 cross-peak. We can also see a cross-peak 2-3 that confirms that the olefinic protons are *cis*-related, since *trans*-related protons are not close enough in space to show an NOE.

Nuclear Overhauser effects are inherently even weaker for molecules with intermediate molecular weights, and are actually zero at a particular molecular weight and a particular viscosity. This problem is overcome to some extent with a slightly different pulse sequence called ROESY (Rotating frame Overhauser Enhancement SpectroscopY). ROESY spectra look similar to NOESY spectra, although actually they differ in that the diagonal spectrum and the cross-peaks in a ROESY are of different sign. This does not affect the way that they are interpreted.

Larger molecules (1000 Daltons or more, in non-viscous solvents, or greater than 500 Daltons in a more viscous solvent such as d_6-DMSO) normally give much more intense NOE cross-peaks, and can give a wealth of information about their geometry. The NOE is caused by mutual dipolar relaxation of protons (Secs. 3.10 and 3.13) and, since this effect falls off with $1/r^{-6}$ (where r is the distance between the protons), the intensity of NOE cross-peaks falls off rapidly with increasing internuclear separation (but is also affected by other variables). A useful guide for larger molecules is: large cross-peaks, r = 2.0-2.5 Å; medium-sized cross-peaks, r = 2.0-3.0 Å; small cross-peaks, r = 2.0-5.0 Å. It is in determination of the solution 3D structures of small proteins that the NOESY spectrum has made its greatest impact upon science.

3.21 2D-TOCSY spectra

In Sec. 3.16, we saw how a 1D-TOCSY spectrum identified the components of a spin system, and separated them from all other signals, both from within the molecule and from impurities. In the case of the ketone **105** it was easy to separate the two AA'MX$_3$ systems by taking one TOCSY spectrum for each. In a larger molecule with a larger number of separate spin systems, this would be tiresome, since we would have to take a 1D spectrum for each. It would also be difficult to irradiate only one signal if a large number were close together in chemical shift. For larger molecules, a 2D version of TOCSY solves these problems, and reveals all the spin systems in one experiment. Fig. 3.64 shows the 2D-TOCSY spectrum of the same molecule, the ketone **105**. Although unnecessary for a small molecule like this, it will keep the picture reasonably simple.

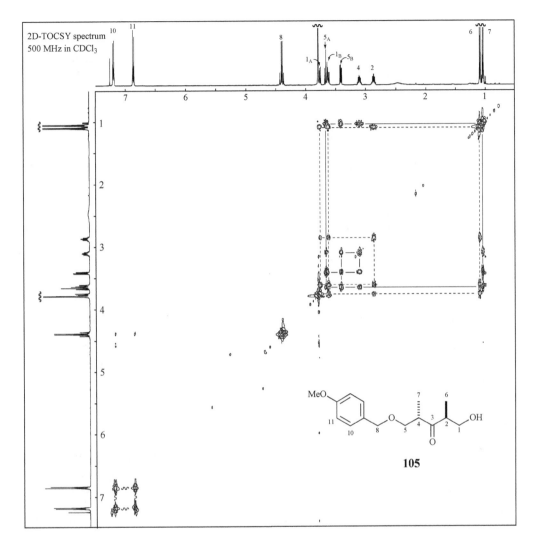

Fig. 3.64

As usual the normal spectrum is on the diagonal, and cross-peaks are found for protons that are part of the same spin system. Thus, the protons on C-4, C-5 and C-7 make up one spin system, connected by the cross-peaks picked out with the solid squares. The protons on C-1, C-2 and C-6 make up another spin system, connected by the cross-peaks picked out with the dashed squares. The lines are close but it is easy enough to see which cross-peaks are on which line. The aromatic protons on C-10 and C-11 make up a third spin system, picked out with the wavy lines. Whereas in a COSY spectrum for this molecule, cross-peaks would be found for the connection between the protons on C-1 and C-2 and between the protons on C-2 and C-6, in the TOCSY spectrum, there is also a cross-peak between the protons on C-1 and C-6. One of the advantages of this technique therefore lies in the interpretation of spectra where there are overlapping resonances. For example, consider the hypothetical case where the M resonance of an AMX system overlaps with the M′ resonance of an A′M′X′ system. From a COSY spectrum, the cross-peaks would not tell us if A was part of the same spin system as X or X'. In the corresponding TOCSY spectrum, the connection from A to X, and from A' to X' would be clear.

TOCSY spectra are important in structural studies of peptides and proteins. The NH protons in amides, unlike the NH protons of amines, frequently show coupling to the adjacent α-CH, and therefore appear as part of the spin system of each component amino acid, but the carbonyl groups insulate the signals of one amino acid from the next. The NH signals are often well spaced, because many of them indulge in more or less hydrogen bonding, which spreads them out on the chemical shift axis. They can often be identified from the TOCSY spectra, first by taking a 2D-TOCSY to get the whole picture, and then by carrying out selective 1D-TOCSY experiments, like those illustrated in Fig. 3.55, to reveal the coupling pattern and the multiplicities that help in identifying the individual amino acids.

Peptides and proteins are water soluble, and it is necessary to remove most or all of the signal from the water. It would not help to use D_2O, because this would exchange with the NH protons that are so useful. The unwanted resonance of H_2O can be largely suppressed by presaturation of its resonance before the acquisition of each TOCSY spectrum, but it does leave some distortion in the 4-5 p.p.m. region.

With only limited chemical information and a molecular weight from FIB or ESI mass spectrometry (Chapter 4), it is possible from complete COSY, NOESY and TOCSY spectra to solve structures such as those of oligopeptides, oligosaccharides and small proteins with three-dimensional detail. A combination of COSY, NOESY and TOCSY spectra does not give quite the wealth of data and precision available from X-ray crystallography. However, these techniques have the compensating advantages that they apply to molecules in solution (in which state biology most commonly functions), and the compound does not have to be crystalline.

3.22 ^{1}H-^{13}C COSY spectra

The 2D spectroscopic techniques that we have seen so far have been from protons to protons. It is also possible to examine the connections from protons to other nuclei, of which the ^{1}H-^{13}C COSY spectra are the most important. If the proton spectrum can be correlated with the carbon spectrum, then the complete assignment of both is easier. Additionally, the carbon dimension can be used to resolve the often severely overlapping proton dimension.

3.22.1 Heteronuclear Multiple Quantum Coherence (HMQC) spectra

One of the ^{1}H-^{13}C COSY experiments allows this correlation to be made by using a pulse sequence in which, following the 'preparation' of the nuclear spins, a delay time in the pulse sequence is set to $1/2J$, where J is the value of the one-bond ^{13}C-^{1}H coupling constant, which is usually in the range 100-200 Hz (see Table 3.14). In this experiment, a correlation is achieved between ^{13}C and the proton to which it is *directly* attached. This is usually called an HMQC spectrum, and an example is provided by that of the ketone **105** in Fig. 3.65, where the proton spectrum is on the horizontal axis at the top and the carbon spectrum is on the vertical axis on the left-hand side.

There is no spectrum on the diagonal, since the HMQC spectrum is effectively a plot of the correlation of chemical shift for protons against carbons—and this is not a good straight line! The peaks that appear in the plot identify the one-bond connections, and so three carbons, C-3 (δ 217.4), C-9 (δ 129.8) and C-12 (δ 159.3), with no protons directly

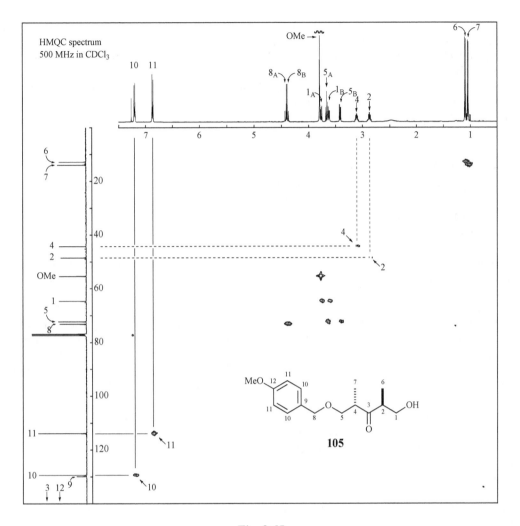

Fig. 3.65

attached, do not give rise to any cross-peaks. The downfield ^{1}H signal at δ 7.21 on the x-axis from the aromatic protons on the two equivalent carbon atoms C-10 is aligned with the downfield carbon signal on the y-axis at δ 129.3, and the upfield signal at δ 6.74 from the aromatic protons on C-11 is aligned with the upfield ^{13}C signal at δ 113.8, as shown by the two horizontal and two vertical solid lines on Fig. 3.65. However, since the proton and the carbon signals do not make a linear plot along the diagonal, it is unsafe to assume that the downfield proton will always correlate with a downfield carbon. For example, the downfield methine proton on C-4 at δ 3.12 correlates with the upfield carbon at δ 44.2, and the upfield methine on C-2 at δ 2.88 with the downfield carbon at δ 48.5, as shown by the two horizontal and two vertical dashed lines on Fig. 3.65. Similarly, the methyl groups C-6 and C-7 are not in the same order on the ^{1}H spectrum as they are on the ^{13}C spectrum. The ChemNMR programme estimating the chemical shifts for this molecule gets most of it right, but not all—the estimation for both the protons and the carbons on C-2 and C-4 are all in the right order, but C-5 and C-8, although in the right order for the protons, are in the wrong order for carbon, and the two methyl groups, C-6 and C-7 are predicted to be identical in the proton spectrum. The lesson is that if you want to assign a ^{13}C spectrum securely, you need to run an HMQC spectrum— prediction programmes are not reliable enough. Another pulse sequence called HSQC gives a very similar 2D spectrum, interpreted in the same way, but shows none of the proton-proton coupling that stretches the peaks out in the horizontal direction. It helps in the interpretation of complex spectra, when peaks often overlap, to have the cross-peaks smaller and more nearly circular.

3.22.2 Heteronuclear Multiple Bond Connectivity (HMBC) spectra

A second ^{1}H-^{13}C COSY pulse sequence gives what is usually called an HMBC spectrum, which also has ^{13}C chemical shifts on one axis and ^{1}H chemical shifts on the other. After the 'preparation' of the nuclei, a time delay in the pulse sequence for an HMBC spectrum is set to correspond to $1/2J$, where J is in the region of 10 Hz, typically about 50 ms. Since many $^2J_{CH}$ and $^3J_{CH}$ coupling constants are in the range 2-20 Hz (Sec. 3.5), the ^{13}C chemical shifts are now correlated with the chemical shifts of those protons *separated from them by two and three bonds*. It is an unfortunate limitation that the values of the two- and three-bond couplings are so similar, and so it is not possible to identify whether a connection, revealed in the HMBC spectrum, is through two or three bonds. The technique is nevertheless powerful, because it is often possible to identify the connection between one spin system and another. In this way, complete connectivity in complex skeletons can often be deduced. To see how it works, we shall look only at a simple example, the ketone **105**. Fig. 3.66 shows the HMBC spectrum of the ketone **105**, where some of the more informative two- and three-bond couplings are labelled, with the latter in brackets.

Although almost all of the peaks in Fig. 3.66, both labelled and unlabelled, are consistent with the two- and three-bond couplings expected for this molecule, there are a few weak signals from one-bond couplings, matching those in the HMQC spectrum. This is frequently the case in HMBC spectra. Key peaks, subtended by the horizontal and vertical lines, are those labelled (C8-H5). They show that C-8, giving the signal at δ 73.1, is two or three bonds from the protons on C-5, giving the signals at δ 3.78 and 3.69. This information is in fact transmitted through three bonds, and permits a structural connection of a previously established proton spin system associated with the aromatic ring to the one extending from the right of the oxygen atom. *Thus, there is communication across the ether linkage that allows, in the case of an unknown, further definition of the skeleton of the molecule.* Equally, the unresolved peaks labelled (C5-H8), picked out with the dashed lines, establish the same connection in the other direction.

The HMBC spectrum also tells us how the fully substituted carbons C-3, C-9 and C-12 are connected to the rest of the molecule. The weak carbon signal, at δ 159.3 from C-12 (previously distinguished from C-9 only by its chemical shift, reliably as it happens) shows HMBC correlations to the proton signals on C-10 and C-11, and, most tellingly, to the methoxy signal at δ 3.79. The other aromatic carbon with no attached protons, C-9 at δ 129.8, has no cross-peak with the methoxy signal. The peaks labelled (C3-H1) and (C3-H5) and those labelled (C3-H4) and (C3-H2) connect the carbonyl group into the carbon framework that extends to both of its sides. *Thus, HMBC allows the extension of skeletal information not only across heteroatoms, but also across carbonyl groups.* The peaks labelled (C8-H10) and (C10-H8) establish the three-bond coupling from C-8 to the protons on C-10.

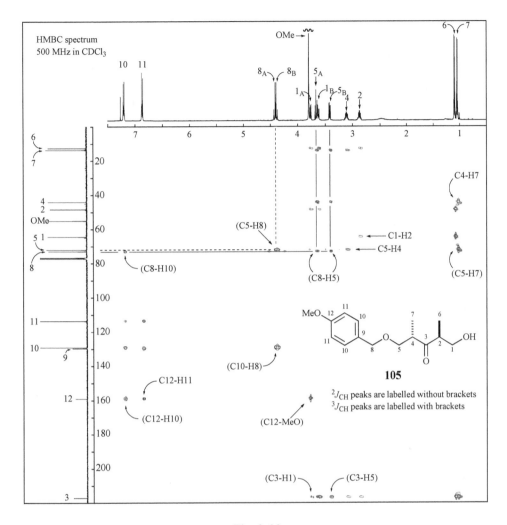

Fig. 3.66

3.23 Measuring ^{13}C-^{1}H coupling constants (HSQC-HECADE)

The HSQC-HECADE (Heteronuclear Single Quantum Coherence) experiment is rather more specialised than the 2D spectra described above. In favourable cases, including those when a methylene group sits between two stereogenic centres, it allows relative configuration to be assigned in open-chain compounds. The HSQC-HECADE experiment measures ^{13}C-^{1}H coupling constants, including the sign, and identifies which proton is coupled to which carbon with which coupling constant. Since coupling constants are dependent upon dihedral angles, they give information about conformation and this in turn can be used to deduce relative configuration.

We saw in Sec. 3.5 that $^{1}J_{CH}$, $^{2}J_{CH}$ and $^{3}J_{CH}$ can be measured in the ^{13}C spectrum when proton decoupling is turned off, but, with more complicated compounds than the ester $\mathbf{1}$, it is not so easy to measure these coupling constants, and it can be impossible to identify the protons to which the ^{2}J and ^{3}J coupling occurs. Determining relative stereochemistry using NMR spectroscopy is a particularly difficult task in open-chain compounds, where conformation is unpredictable and NOEs are unreliable.

As an example to illustrate how HSQC-HECADE spectra are read, we shall use the diol **109**, which was prepared by the addition of a propynyl nucleophile to the meso dialdehyde **108**, followed by separation of the four products. Two of the products, including the isomer **109**, were identifiably centrosymmetric by their having only one set of signals for each half. The problem was to determine which was which. The diol **109** turned out to be the anti isomer with respect to the relative configuration between C-4 and C-6, as we shall see. COSY and HMQC spectra had allowed all the signals to be assigned for both centrosymmetric isomers, and the vicinal and geminal proton-proton coupling constants to be measured. HSQC-HECADE spectra then gave all the $^2J_{CH}$ and $^3J_{CH}$ values, and these in turn allowed the relative stereochemistry between C-4 and C-6 to be determined. Part of the HSQC-HECADE spectrum for the anti isomer is shown in Fig. 3.67, which resembles an HMQC spectrum, with the cross-peaks connecting the proton spectrum at the top with the ^{13}C spectrum on the left-hand side.

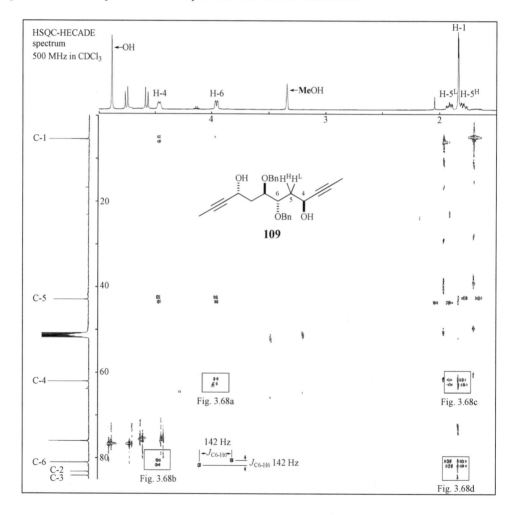

Fig. 3.67

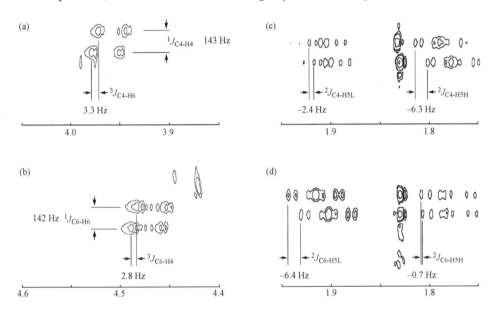

Let us first look at the cross-peaks between C-6 (at δ 81.05) and the double-doublet for H-6 (at δ 3.96). In the HMQC spectrum this is a single peak identifying only that these two atoms are connected, but in the HSQC-HECADE spectrum in Fig. 3.67 it is a pair of peaks, separated in both dimensions by 142 Hz. This is the $^1J_{CH}$ coupling constant between these two atoms. This is not useful information for structure determination, but it illustrates how HSQC-HECADE spectra are read. For more detailed use of the HSQC-HECADE spectrum, we look at the more telling expansions in Fig. 3.68.

Fig. 3.68

The $^3J_{CH}$ values are measurable in the same way as the $^1J_{CH}$ values—they can be seen more clearly in Figs. 3.68a and 3.68b—with the $^3J_{CH}$ values of 3.3 Hz for $^3J_{C4\text{-}H6}$ and 2.8 Hz for $^3J_{C6\text{-}H4}$ represented by the horizontal displacement. These numbers are probably only accurate to the nearest whole number. Being at the lower end of the range (0-8) for $^3J_{CH}$ values, they suggest that the carbon atoms and the hydrogens are gauche rather than antiperiplanar. At this stage it is not known whether this isomer has the anti, or the syn, relationship between the stereocentres at C-6 and C-4. Therefore, the conformation about the C4-C5 bond could be a consequence of any relative weighting of the conformations **110** and **111**, or of **112** and **113**. Similarly, the conformation about the C5-C6 bond could be a consequence of any relative weighting of the conformations **114** and **115**. The $^1J_{CH}$ values for C-4 and C-6 of 143 and 142 Hz, respectively, are easily measurable by the vertical displacements in Figs. 3.68a and 3.68b, but they are not needed in this structure determination.

110 **111** **112** **113**

114 **115**

The next step in the analysis is to assign which is which of the two methylene protons on C-5, labelled H-5^L for the low-field signal at δ 1.91 and H-5^H for the high-field signal at δ 1.79. The vicinal proton-proton coupling constants are 3.3 Hz between H-6 and H-5^H, 9.3 Hz between H-6 and H-5^L, 9.3 Hz between H-4 and H-5^H, and 4.0 Hz between H-4 and H-5^L. These are governed by a Karplus relationship in the usual way, and indicate that H-5^L is predominantly anti to H-6 and gauche to H-4, whereas H-5^H is predominantly gauche to H-6 and anti to H-4, indicating that the conformations **110-115** can be labelled with a little more detail as any of the possibilities **110a-115a**.

110a **111a** **112a** **113a**

114a **115a**

To decide between the various possibilities, we need the $^2J_{CH}$ coupling constants which are found in the detail of Figs. 3.68c and 3.68d. The $^2J_{CH}$ displacements are in the opposite direction from the $^1J_{CH}$ and $^3J_{CH}$ displacements, because $^2J_{CH}$ coupling constants are negative. For the proton H-5^L resonating at low field, the $^2J_{CH}$ values are, to the nearest whole number, −2 Hz to C-4 and −6 Hz to C-6, and for the proton 5-H^H resonating at high field, they are −6 and −1 Hz. These numbers are not governed by the Karplus equation as the $^3J_{CH}$ values are. They are governed by the relationships of the carbon and proton to the other substituents, especially when one of them is electronegative, as here. Assigning the conformational relationships from them uses a large amount of empirical data, which suggests, in this case, that the relationships in **111a**, **112a** and **114a**, in which the oxygen substituent is anti to the proton with the smaller $^2J_{CH}$ coupling constant, are more compatible with precedent than are the other relationships. The full protocol used to identify which of the possibilities like those shown as **110a-115a** is true is given for the general case in a paper by Murata (N. Matsumori, D. Kaneno, M. Murata, H. Nakamura and K. Tachibana, *J. Am. Chem. Soc.*, 1999, **64**, 866-876). The result in this case is the unambiguous assignment that the conformation **114a** is the more populated of the pair **114a** and **115a**. This shows that H^H is the hydrogen atom which would project towards the reader in the drawing **109** and H^L is the hydrogen atom which would project away from the reader; this in turn shows that the configuration **111a** is correct rather than **112a**. Putting together the two drawings **111a** and **114a** defines the preferred configurational relationship between C-4 and C-6 as anti. Furthermore, we now know that the most populated conformation of this half of the molecule is that shown in Fig. 3.69, which includes the coupling information that was used to arrive at it. The other half of the centrosymmetric molecule, of course, exactly mirrors this half.

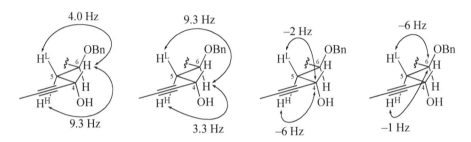

Fig. 3.69

3.24 Identifying ^{13}C-^{13}C connections (INADEQUATE spectra)

The spectra discussed in Secs. 3.21 and 3.22 show us how to make connections from ^{13}C to ^1H through one, two and three bonds, but it is also useful to have a technique to show direct ^{13}C-^{13}C connections. A pulse sequence called INADEQUATE (Incredible Natural Abundance DoublE QUAnTum coherencE) does this, but there is a problem of sensitivity: the natural abundance of ^{13}C is close to 1%, making the signals inherently weak, but, in addition, the probability of finding one ^{13}C attached to another is even lower, only 1 in 10 000. Unless the molecule is enriched in the ^{13}C isotope, an INADEQUATE spectrum can only be taken with a large amount of material in relatively concentrated solution, and even so with long runs. It is not therefore a routine experiment, but as instruments grow ever more sensitive, it is a powerful technique that may well increase in importance.

Figure 3.70 shows the INADEQUATE spectrum of the ester-aldehyde **116**, together with the conventional ^{13}C spectrum on each axis. It required 250 mg of sample, a cryoprobe, a 500 MHz spectrometer and 48 hours of acquisition time. It also used an updated version of the INADEQUATE protocol which makes the spectrum appear rather more like a COSY spectrum than earlier versions, and thus easier to interpret.

Whereas COSY spectra show $^2J_{HH}$, $^3J_{HH}$ and $^4J_{HH}$ coupling, with the proton spectrum on the diagonal, and the cross-peaks reflected across the diagonal, the INADEQUATE spectrum shows $^1J_{CC}$ coupling, and has neither the full ^{13}C spectrum on the diagonal nor perfect reflection symmetry across the diagonal. In the case of the ester **116**, three of the carbon atoms do in fact show up on the diagonal, namely the peaks labelled α, β and γ, which are the signals from C-9, C-1 and C-2', respectively. Each of these atoms is at the terminus of a chain, and that appears to influence whether or not the peak will appear on the diagonal.

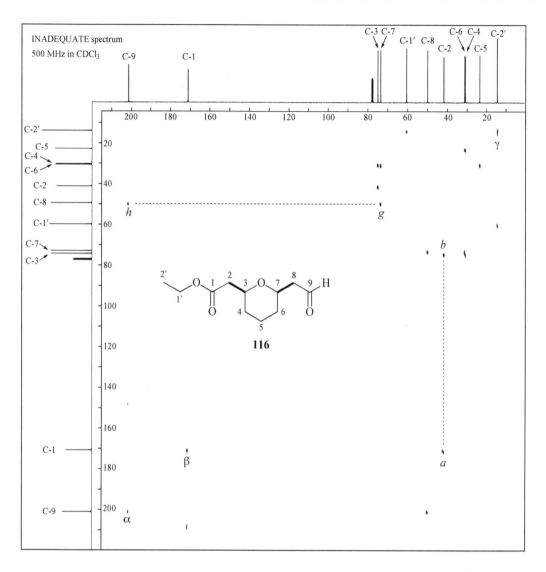

Fig. 3.70

It is best to start with a signal which can be assigned unambiguously, such as the signal of C-1, from the carbonyl group of the ester, which appears, typically downfield, at δ 170.8. Working from C-1, the cross-peak labelled *a* is horizontally aligned with the signal from C-1 on the vertical scale at the left-hand side of Fig. 3.70 and it is directly below the signal at δ 41.1 on the horizontal scale at the top. This tells us that the signal at δ 41.1 is produced by C-2. The cross-peak *a* is, in turn, on the same vertical line (dashed) as the cross-peak *b*, which is horizontally aligned with the signal at δ 74.3 on the vertical axis, showing that the signal at δ 74.3 must be from C-3. To continue the assignment, it is helpful to look at the enlargement of the high-field end of the spectrum shown in Fig. 3.71.

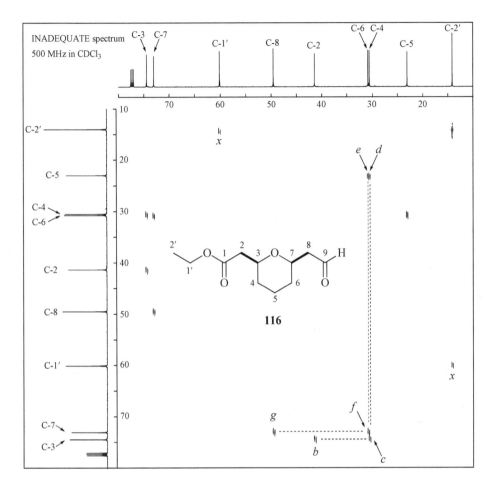

Fig. 3.71

Continuing the analysis using Fig. 3.71, we have reached the cross-peak labelled *b*, which is connected by the horizontal line to the cross-peak *c*. This cross-peak is directly below the signal at δ 30.1, which must be from C-4. Following the vertical dashed line from *c* to the cross-peak *d* then identifies the connection from C-4 to C-5, which we can see gives rise to the peak at δ 22.9. From *d*, there is no obvious connection, and so we deduce that *d* must connect to the signal *e*, which is so close to *d* that we cannot easily see its fine structure. The cross-peak *e* is vertically below the signal at δ 30.7, which must be the signal from C-6, not surprisingly very close in chemical shift to the signal from C-4. The cross-peak *e* is in turn vertically above the cross-peak *f*, which is aligned with the signal at δ 73.1 on the vertical axis, and which must therefore be from C-7. The cross-peak *f* is connected horizontally to the cross-peak *g*, vertically below the signal at δ 49.5 on the horizontal axis, which is the signal from C-8. Finally, returning to the full spectrum in Fig. 3.70, we can see that the cross-peak *g* is connected to the cross-peak *h*, which is vertically below the signal at δ 201.1, the signal from the aldehyde carbonyl carbon, C-9. The cross-peaks *x* (Fig. 3.71), which do not connect on to any other cross-peaks, identify the connection between C-1′ and C-2′ in the ethyl group. In summary, the cross-peaks *a-h* and *x* make the connections shown in the drawing **117**.

117

Note that, as we make each connection, we change direction through 90° as we look for the next cross-peak. We could equally easily have started from the aldehyde carbonyl group, and followed the chain of connections right through to C-1, using the cross-peaks not labelled in Figs. 3.70 and 3.71.

Fig. 3.71 also reveals that the cross-peaks are all doublets, because the signals detected are from one ^{13}C attached to only one ^{13}C. This situation arises because the probability of any ^{13}C being attached to more than one carbon is another factor of 100 less than the already low probability of 1 in 10 000. The separation of the doublets measures the coupling constant, $^{1}J_{CC}$. In this case, fairly typically, all the coupling constants are in the range 30-40 Hz, but in some cases they can be as high as 70 Hz.

3.25 Three- and four-dimensional NMR

When the signals of an NMR spectrum are spread out into a second dimension (Sec. 3.17), the dispersion of the spectroscopic information is thereby improved. It would clearly be an advantage to extend this concept by obtaining three- (and even four-) dimensional NMR spectra. Such spectra are in fact available, and although only a very limited treatment of this rather specialised subject is justified here, 3D and 4D spectra are powerful in determining the three-dimensional structures of relatively large and important molecules such as polypeptides, proteins and lengths of DNA duplexes (DNA double helices). This area is important in the realm of structural biology, but is covered only in very brief outline here.

Consider the problem of determining the three-dimensional structure of a small protein, which typically might consist of a sequence of 100 amino acids, where the segment **118** shown has arbitrarily an alanine and a serine residue, which may or may not be adjacent.

118

The first step would be to identify the proton spin systems of each amino acid, which can in principle be achieved by using 2D TOCSY spectra (Sec. 3.20), but in practice too many of the signals are likely to overlap in a large molecule. For example, if the N*H* of the alanine residue and the N*H* of the serine residue in the protein, whether they are adjacent or not, had the same proton chemical shift, the proton spin systems of the two amino acids would both have cross-peaks on a line defined by the N*H* chemical shift in the TOCSY spectrum, which might look like Fig. 3.72a. The problem arising because the two NHs have the same proton chemical shift can be overcome by carrying out the biosynthesis of the protein in an environment where it is given ^{15}N-enriched ammonium nitrate as its nitrogen source. The protein is then produced as its ^{15}N-enriched analogue, and it is possible to take a 3D TOCSY spectrum so that each of the 2D TOCSY spectra are separated into a third dimension. Each TOCSY spectrum is found in a plane defined by the ^{1}H axes in the usual way, but the planes are separated according to the ^{15}N chemical shift of each amino acid. If the alanine N and the serine N have different chemical shifts, the two planes might look something like Fig. 3.72b, where the TOCSY signals from the alanine and the valine are separated onto different planes. As the name for this experiment implies, 3D ^{15}N-TOCSY-HMQC, these spectra can be regarded as the combination of a 2D TOCSY, with an HMQC in the third dimension, except that in this case the heteronuclear component is ^{15}N and not ^{13}C.

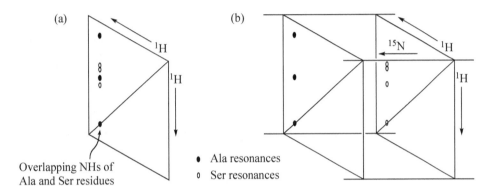

Fig. 3.72

Once the spin systems of the amino acids of the protein are assigned, all the NOEs which are available between the assigned proton resonances must be analysed. The 3D experiment for this assignment is the 3D ^{15}N-NOESY-HMQC spectrum. In this spectrum the planes that show the TOCSY cross-peaks in Fig. 3.72b now have the NOESY cross-peaks. Thus, as long as the NH proton resonance of each amino acid is defined by a unique combination of ^{15}NH and NH chemical shifts, all the protons near in space to the NH proton of a specified amino acid can be identified. This is particularly useful in defining which parts of the protein form α-helices and which form β-sheets. In an α-helix **119**, the NH of the ith amino acid will give an NOE to the NH of the (i+1)th, whereas in a β-sheet **120** it is the α-CH of the ith amino acid which gives an NOE to the NH of the (i + 1)th amino acid. Using these NOEs, and others which will not be detailed here, further detail of the secondary structure in solution can often be assigned.

Finally, other through-space interactions can be picked up using a third 3D experiment giving a 3D TOCSY-NOESY spectrum, in which a series of 2D TOCSY spectra are created with each collecting all the NOESY connections to a single identifiable proton. The diastereotopic methyl groups of valine and leucine, for example, are apt to give rise to recognisable sharp doublets, and to be held in space in a position to interact through space with other residues. Thus, a series of 3D spectra can identify the signals from the component amino acids, and then establish how the amino acid residues, which may be

119

Two amino acids in the conformation of an α-helix: NOEs from NH_i to NH_{i+1} are seen.

120

The conformation of a β-sheet: hydrogen bonds are indicated by dashed lines. NOEs are seen from the α-CH of the ith to the NH of the ($i+1$)th amino acid

distant from each other in the linear sequence, can be close in the folded structure. Given enough NOEs and a good enough separation of the ^{15}NH and NH signals, the full folded structure of a protein can be described with a precision approaching that found by X-ray crystallography. Such information has the further advantage over most X-ray structures that it gives details of where the hydrogen atoms are, and it applies to the solution structure.

3.26 Hints for spectroscopic interpretation and structure determination

The simple ^1H and ^{13}C NMR spectra, with a wealth of information in the chemical shift and coupling constants, will take you a long way towards determining the structure of an unknown compound, or confirming the structure of the product of a reaction. But if they are not sufficient on their own, one or more of the special pulse techniques described in this chapter will almost certainly give you all the information needed. The APT spectrum, simple spin decoupling, difference decoupling, COSY, NOESY, TOCSY, HMQC and HMBC spectra, are the tools that most often complete the picture.

3.26.1 Carbon spectra

1. Check the peaks caused by the solvent (Table 3.26).
2. Check whether the number of ^{13}C signals is consistent with the molecular formula, if one is available from HRMS (Chapter 4). If it is larger than that required by the molecular formula, be aware that there may be weak signals from traces of impurities. may be comparable in intensity with. If, on the other hand, the number of ^{13}C peaks is smaller than that required by the molecular formula, look for weak signals from carbon atoms having no attached protons, and consider the possibility that symmetry within the molecule has led to duplication of some or all of the signals.
3. Using an APT (or DEPT if necessary) spectrum, determine the relative numbers of CH_3, CH_2, CH and fully substituted carbons.
4. Attempt to estimate the relative numbers of carbonyl and other trigonal and tetrahedral carbons, noting whether the latter are in aliphatic or electronegative environments.

3.26.2 Proton spectra

1. Check for the peaks caused by the solvent (Table 3.26).

2. Search for a proton signal that is plausibly caused by one proton (or, less desirably, two or three protons). From this, using the integrals, make a best possible estimate of the number of protons in each signal and in the spectrum as a whole, and check if this estimate is consistent with other evidence, for example from a molecular weight determined by the mass spectrum or, better still, from an exact mass measurement.

3. Check for protons that can be exchanged for deuterium (e.g. NH or OH). These may be broad, or temperature or pH dependent; at some stage in the analysis, you may wish to ascertain whether these postulated resonances disappear following a 'D$_2$O shake'.

4. Estimate the numbers of protons in aliphatic, electronegative, and trigonal (sp^2) environments. When dealing with aliphatic environments, note the signals given by primary, secondary, or tertiary methyl groups, and be aware that signals given by vinyl methyl resonances are typically somewhat broader than those associated with tertiary methyl groups.

5. Analyse any resolved coupling patterns, using the magnitudes of coupling constants. If necessary, run specific decoupling experiments, or obtain the COSY spectrum. This is often the biggest and most enjoyable stage of the quest for structural information; practice, and a familiarity with coupling patterns and roofing, is a great help.

6. If information on the spatial relationships between protons is required (including conformational and configurational information) make use of the Karplus equation and NOEs.

7. A 'non-invasive' way to determine the number of hydroxyl protons (and their multiplicity) in a molecule is to determine the spectrum in d$_6$-DMSO under conditions that are mildly acidic ('pH' in the range 2-4). This solvent usually contains traces of water which resonate near δ 3.4. The normal spectrum (A) is recorded and stored in the spectrometer's computer memory, and a second spectrum (B) is similarly stored, but this time obtained with simultaneous irradiation of the water signal. In the latter spectrum, the OH protons have been removed by transfer of saturation from the water signal. The readily displayed difference spectrum (A – B) therefore contains only the signals from the water and the various OH groups (which, at 'neutral pH', are in slow exchange with each other and with the water peak). Note that the disadvantage of DMSO as a solvent is that it is subsequently not easy to remove it completely from the sample.

3.26.3 Hetero-correlations

1. If necessary, identify spin systems using a 1D or 2D TOCSY spectrum.

2. Correlate the carbon and proton assignments using an HMQC spectrum (^{1}H-^{13}C COSY, Sec. 3.22.1).

3. In conjunction with the information from 2, use an HMBC spectrum (long-range ^{1}H-^{13}C COSY, Sec. 3.22.2) to attach identified CH$_3$, CH$_2$ and CH groups to quaternary and carbonyl carbons.

4. As a last resort, rarely necessary and only if you have enough compound, take an INADEQUATE spectrum to identify the connectivity of the carbon atoms.

3.27 Internet

The Internet is an evolving system, with links and protocols changing frequently. The following information is inevitably incomplete and may no longer apply, but it gives you a guide to what you can expect. Some websites require particular operating systems and may only work with a limited range of browsers, many require payment, and some require you to register and to download programs before you can use them.

For guides to spectroscopic data on the Internet, see the websites at MIT, the University of Waterloo and the University of Texas, representative of several others. They are tailored for internal use, but are informative nevertheless:

http://libraries.mit.edu/guides/subjects/chemistry/spectra_resources.htm
http://lib.uwaterloo.ca/discipline/chem/spectral_data.html
http://www.lib.utexas.edu/chem/info/spectra.html

There are vast resources on the Internet, for ^{1}H and ^{13}C data. Several of these (CDS, Bio-Rad, ACD, Sigma-Aldrich, ChemGate) cover most or all of the spectroscopic methods, and have already been mentioned in the corresponding section in the chapter on IR on pp. 42-43. Many websites, especially those reached by way of Google, will carry advertisements to commercial organisations measuring and interpreting NMR spectra.

A website leading to software for NMR processing, visualisation, assignment and prediction is:

http://www.netsci.org/Resources/Software/Struct/nmr.html
This includes such programs as: ACD/NMR SpecManager, Acorn NMR, Aurelia, Azarea, Felix (Accelrys), Gifa, gNMR, HyperNMR, JspecView, KnowItAll (Sadtler), MestRe-C 1.0, NMR Refine (Accelrys), Perch, UXNMR (Bruker).

A site listing databases for IR, NMR and MS is:
http://www.lohninger.com/spectroscopy/dball.html

The Japanese Spectral Database for Organic Compounds (SDBS) at:
http://www.aist.go.jp/RIODB/SDBS/cgi-bin/cre_index.cgi
has free access to IR, Raman, ^{1}H- and ^{13}C-NMR and MS data.

There are several servers in Germany giving access to the NMRShiftDB facility:
http://nmrshiftdb.ice.mpg.de/
for searches of over 20 000 ^{1}H and ^{13}C NMR spectra, continually being added to, together with the capacity for making predictions.

The Sadtler database administered by Bio-Rad Laboratories has, for a price, over 360 000 ^{13}C and 30 000 ^{1}H NMR spectra of pure organic and commercial compounds. For information, go to:
http://www.bio-rad.com/
and follow the leads to Sadtler, KnowItAll, and NMR.

The Wiley-VCH website gives access to the SpecInfo data:
http://www3.interscience.wiley.com/cgi-bin/mrwhome/109609148/HOME
and to ChemGate, which has a collection of 700 000 IR, NMR and mass spectra:
http://chemgate.emolecules.com

ChemDraw, containing ChemNMR for chemical shift predictions, is produced by Cambridge Soft: www.cambridgesoft.com/

3.28 Bibliography

HOW NMR WORKS
A. E. Derome, *Modern NMR Techniques*, Pergamon, Oxford, 1987.
R. R. Ernst, G. Bodenhausen and A. Wokaun, *Principles of Nuclear Magnetic Resonance in One and Two Dimensions*, OUP, Oxford, 1987.

J. K. M Sanders and B. K. Hunter, *Modern NMR Spectroscopy: A Guide for Chemists*, 2nd Ed., OUP, Oxford, 1993.

P. J. Hore, *Nuclear Magnetic Resonance*, OUP, Oxford, 1995.

P. Hore and J. Jones, *NMR: The Toolkit*, OUP, Oxford, 2000.

D. M. Grant and R. K. Harris (Eds.), *Encyclopedia of Nuclear Magnetic Resonance*, 8 Vols., Wiley-VCH, Weinheim, 1996.

R. Freeman, *A Handbook of Nuclear Magnetic Resonance*, Longman, 2nd Ed., 1997.

J. Keeler, *Understanding NMR Spectroscopy*, Wiley, Chichester, 2005.

STRUCTURE DETERMINATION

R. J. Abraham, J. Fisher and P. Loftus, *Introduction to NMR Spectroscopy*, Wiley, New, York, 1988.

J. K. M. Sanders, E. C. Constable and B. K. Hunter, *Modern NMR Spectroscopy: A Workbook of Chemical Problems*, 2nd Ed., OUP, Oxford, 1993.

S. Berger and S. Braun, *200 and More NMR Experiments: A Practical Course*, Wiley-VCH, Weinheim, 2004.

E. Breitmaier, *Structure Elucidation by NMR in Organic Chemistry*, Wiley, New York, 1993.

J. W. Akitt and B. E. Mann, *NMR and Chemistry*, 4th Ed., CRC Press, Boca Raton, 2000.

T. N. Mitchell and B. Costisella, *NMR—From Spectra to Structures*, Springer, Heidelberg, 2004.

H. Friebolin, *Basic One- and Two-Dimensional NMR Spectroscopy*, 4th Ed., Wiley-VCH, Weinheim, 2005.

DATA

C. J. Pouchert, *The Aldrich Library of ^{13}C and ^{1}H FT NMR Spectra*, 3 Vols., Aldrich Chemical Company Inc., 1993; also available on CD.

E. Pretsch, P. Buhlmann and C. Affolter, *Tables of Spectral Data for Structure Determination of Organic Compounds*, 3rd English Ed., Springer, Berlin, 2000.

T. J. Bruno and P. D. N. Svoronos, *CRC Handbook of Fundamental Spectroscopic Correlation Charts*, CRC Press, Boca Raton, 2006.

SPECIALISED ASPECTS OF NMR

K. Wüthrich, *NMR of Proteins and Nucleic Acids*, Wiley, New York, 1986.

G. C. K. Roberts (Ed.), *NMR of Macromolecules*, OUP, Oxford, 1993.

W. R. Croasmun and R. M. K. Carlson (Eds.), *Two-dimensional NMR Spectroscopy—Applications for Chemists and Biochemists*, 2nd Ed., VCH, Weinheim, 1994.

H. Oschkinat, T. Müller and T. Dieckmann, *Protein structure determination with three- and four-dimensional NMR spectroscopy*, in *Angew. Chem., Int. Ed. Engl.*, 1994, **33**, 277-293.

J. A. Iggo, *NMR Spectroscopy in Inorganic Chemistry*, OUP, Oxford, 2000.

D. Neuhaus and M. P. Williamson, *The Nuclear Overhauser Effect*, 2nd Ed., Wiley-VCH, New York, 2000.

J. Jimenez-Barbero, and T. Peters (Eds.), *NMR Spectroscopy of Glyconconjugates*, Wiley, New York, 2002.

Q. T. Pham, R. Pétiaud, H. Waton and M.-F. Llauro-Darricades, *Proton and Carbon NMR Spectra of Polymers*, Wiley, XXX, 2002.

M. Knaupp, M. Bühl and V. G. Malkin (Eds.), *Calculation of NMR and EPR Parameters*, Wiley, New York, 2004.

3.29 Tables of data

Table 3.4 Some parameters of magnetic nuclei

Isotope	NMR frequency (MHz) at 9.396T	Natural abundance (%)	Relative sensitivity	Spin (I)[†]
^{1}H	400.00	99.98	1.00	$^1/_2$
^{2}H	61.40	0.015	0.00965	1
^{3}H	426.80	0	1.21	$^1/_2$
^{7}Li	155.44	92.58	0.293	$^3/_2$
^{11}B	128.32	80.42	0.165	$^3/_2$
^{13}C	100.56	1.11	0.0159	$^1/_2$
^{14}N	28.92	99.63	0.00101	1
^{15}N	40.52	0.37	0.00104	$-^1/_2$
^{17}O	54.24	0.037	0.0291	$-^5/_2$
^{19}F	376.32	100	0.833	$^1/_2$
^{23}Na	105.80	100	0.0925	$^3/_2$
^{27}Al	104.24	100	0.206	$^5/_2$
^{29}Si	79.44	4.70	0.00784	$-^1/_2$
^{31}P	161.92	100	0.0663	$^1/_2$
^{35}Cl	39.20	75.53	0.0047	$^3/_2$
^{39}K	18.68	93.10	0.000508	$^3/_2$
^{41}K	10.24	6.88	0.000084	$^3/_2$
^{51}V	105.12	99.76	0.382	$^7/_2$
^{53}Cr	22.60	9.55	0.000903	$^3/_2$
^{55}Mn	98.64	100	0.175	$^5/_2$
^{57}Fe	12.92	2.19	0.000034	$^1/_2$
^{59}Co	94.44	100	0.277	$^7/_2$
^{65}Cu	113.60	30.91	0.114	$^3/_2$
^{79}Br	100.20	50.54	0.0786	$^3/_2$
^{81}Br	108.00	49.46	0.0985	$^3/_2$
^{85}Rb	38.60	72.15	0.0105	$^5/_2$
^{113}Cd	88.72	12.26	0.0109	$^1/_2$
^{119}Sn	149.08	8.58	0.0518	$-^1/_2$
^{133}Cs	52.48	100	0.0474	$^7/_2$
^{195}Pt	86.00	33.80	0.00994	$^1/_2$
^{207}Pb	83.68	22.60	0.00916	$^1/_2$

[†]A minus sign in the Spin column signifies that the magnetogyric ratio γ, and hence the magnetic moment, is negative.

Table 3.5 ^{13}C Chemical shifts in some alkanes

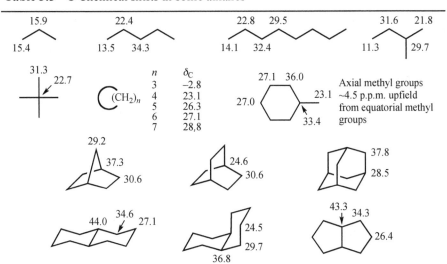

Estimation of ^{13}C chemical shifts in aliphatic chains

$$\delta_C = -2.3 + \Sigma z + \Sigma S + \Sigma K \tag{3.16}$$

where −2.13 is the ^{13}C chemical shift for methane, z is the substituent constant (Table 3.6), S is a 'steric' correction (Table 3.7), and K is a conformational increment for γ-substituents (Table 3.8).

Table 3.6 Substituent constants z for Eq. 3.16

	Substituent	z			
		α	β	γ	δ
H	H—	0	0	0	0
C	alkyl—	9.1	9.4	−2.5	0.3
	—C=C—	19.5	6.9	−2.1	0.4
	—C≡C—	4.4	5.6	−3.4	−0.6
	Ph—	22.1	9.3	−2.6	0.3
	OHC—	29.9	−0.6	−2.7	0.0
	—CO—	22.5	3.0	−3.0	0.0
	—O₂C—	22.6	2.0	−2.8	0.0
	N≡C—	3.1	2.4	−3.3	−0.5
N	R₂N—	28.3	11.3	−5.1	0.0
	O₂N—	61.6	3.1	−4.6	−1.0
O	—O—	49.0	10.1	−6.2	0.0
	—COO—	56.5	6.5	−6.0	0.0
Hal	F—	70.1	7.8	−6.8	0.0
	Cl—	31.0	10.0	−5.1	−0.5
	Br—	18.9	11.0	−3.8	−0.7
	I—	−7.2	10.9	−1.5	−0.9
S	—S—	10.6	11.4	−3.6	−0.4
	—SO—	31.1	9.0	−3.5	0.0

Table 3.7 'Steric' constants S for Eq. 3.16

Observed ^{13}C atom	Number of substituents other than H on the atoms directly bonded to the observed $^{13}C^{\dagger}$			
	1	2	3	4
Primary	0.0	0.0	−1.1	−3.4
Secondary	0.0	0.0	−2.5	−7.5
Tertiary	0.0	−3.7	−9.5	−15.0
Quaternary	−1.5	−8.4	−15.0	−25.0

†Except that CO_2H, CO_2R and NO_2 groups are counted as primary (column 1), Ph, CHO, $CONH_2$, CH_2OH and CH_2NH_2 groups as secondary (column 2), and COR groups as tertiary (column 3).

Table 3.8 Conformational correction K for γ substituents in Eq. 3.16

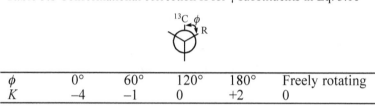

ϕ	0°	60°	120°	180°	Freely rotating
K	−4	−1	0	+2	0

Example of application of Eq. 3.16

Diethyl butylmalonate has ^{13}C signals at δ 13.81, 14.10, 22.4, 28.5, 29.5, 52.03, 61.12 and 169.32.

Take the methine carbon a:

Base value	−2.3	(methane)
1 α-alkyl group	9.1	(carbon b)
1 β-alkyl group	9.4	(carbon c)
3 γ-alkyl groups	−7.5	(carbon d and two Et groups of the OEt groups)
3 δ-alkyl groups	0.9	(Me and two Me groups of the OEt groups)
2 α-CO_2R groups	45.2	
S	−3.7	(a is a tertiary carbon bonded to two CO_2Et groups, which count as primary)
K	0	(open-chain compound with free rotation)
Calculated shift	51.5	Observed value 52.03

There is no difficulty in assigning this signal, because it is the only methine and is easily identified as such. However, the three methylenes at 22.4, 28.5 and 29.5 are less securely identifiable.

The corresponding calculation for carbon *b* is:

Base value	−2.3	(methane)
2 α-alkyl group	18.2	[carbon *c* and (EtO$_2$C)$_2$CH]
1 β-alkyl group	9.4	(carbon *d*)
2 β-CO$_2$R groups	4.0	
1 γ-alkyl groups	−2.5	(Me)
2 δ-alkyl groups	0.6	(CH$_2$ groups of the OEt groups)
2 β-groups	4.0	
S	−2.5	(*b* is secondary and bonded to a carbon, namely *a*, with three groups on it other than hydrogen)
Calculated shift	28.9	

Similar calculations for carbons *c* and *d* give calculated values of 29.1 and 22.8. It is therefore likely, although not certain, that the signals at 22.4, 28.5 and 29.5 can be assigned to *d*, *b* and *c*, respectively. If you have access to a program for predicting chemical shifts you might do a little better. ChemNMR in ChemDraw®, using very similar protocols to Eq. 3.16 and the data in Tables 3.6-3.8, gives:

22.4   28.7
14.1   29.5   CO$_2$Et   52.0   CO$_2$Et

Table 3.9 ^{13}C Chemical shifts in some alkenes, alkynes and nitriles

123.3   31.4   137.2

23.3   32.8   130.8

23.0   25.4   127.4

32.7   28.0   29.6   132.7

26.1   29.2   25.5   130.1

48.8   42.0   135.8   24.8

116.3   136.9

74.8   213.5

R—≡N   110-125

71.9

79.2   —Me   66.9   2.2

84.6   —Ph   78.3

83.0   OH   73.8

Estimation of ^{13}C chemical shifts in substituted alkenes

$$\delta_C = 123.3 + \Sigma z_1 + \Sigma z_2 + \Sigma S \qquad (3.17)$$

where 123.3 is the ^{13}C chemical shift for ethene, z_1 and z_1 are the substituent constants (Table 3.10) and S is a 'steric' correction for alkyl substituents:

For each pair of *cis* substituents	$S = -1.1$
For a pair of geminal substituents on C-1	$S = -4.8$
For a pair of geminal substituents on C-2	$S = 2.5$

Table 3.10 Substituent constants z for Eq. 3.17

	Substituent R	z_1	z_2
H	H—	0	0
C	Me—	10.6	−7.9
	Et—	15.5	−9.7
	Pr^n—	14.0	−8.2
	Pr^i—	20.4	−11.5
	Bu^t—	25.3	−13.3
	ClCH$_2$—	10.2	−6.0
	HOCH$_2$—	14.2	−8.4
	Me$_3$SiCH$_2$—	12.5	−12.5
	CH$_2$=CH—	13.6	−7.0
	Ph—	12.5	−11.0
	OHC—	13.1	12.7
	RCO—	15.0	5.8
	RO$_2$C—	6.3	7.0
	N≡C—	−15.1	14.2
N	RAcN—	6.5	−29.2
O	RO—	29.0	−39.0
	AcO—	18.4	−26.7
Hal	F—	24.9	−34.3
	Cl—	2.6	−6.1
	Br—	−7.9	−1.4
	I—	−38.1	7.0
Si	Me$_3$Si—	16.9	16.1
P	Ph$_2$P(=O)—	8.0	11.0
S	RS—	18.0	−16.0

Example of application of Eq. 3.17

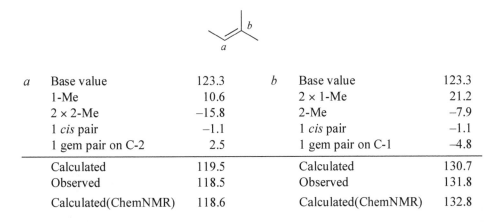

a	Base value	123.3	*b*	Base value	123.3
	1-Me	10.6		2 × 1-Me	21.2
	2 × 2-Me	−15.8		2-Me	−7.9
	1 *cis* pair	−1.1		1 *cis* pair	−1.1
	1 gem pair on C-2	2.5		1 gem pair on C-1	−4.8
	Calculated	119.5		Calculated	130.7
	Observed	118.5		Observed	131.8
	Calculated(ChemNMR)	118.6		Calculated(ChemNMR)	132.8

Table 3.11 ^{13}C Chemical shifts in some arenes

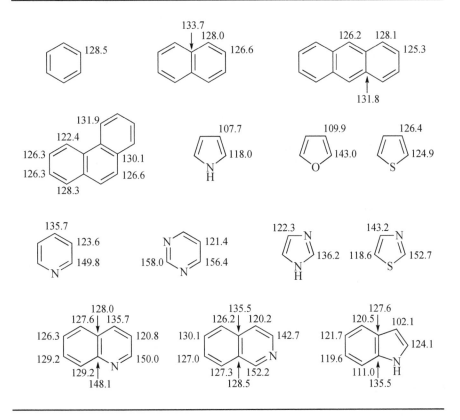

Estimation of ^{13}C chemical shifts in substituted benzenes

$$\delta_C = 128.5 + \Sigma z_i \qquad (3.18)$$

Table 3.12 Substituent constants z_i for Eq. 3.18

	Substituent R	z_1	z_2	z_3	z_4
H	H—	0	0	0	0
C	Me—	9.3	0.6	0.0	−3.1
	Et—	15.7	−0.6	−0.1	−2.8
	Pr^n—	14.2	−0.2	−0.2	−2.8
	Pr^i—	20.1	−2.0	0.0	−2.5
	Bu^t—	22.1	−3.4	−0.4	−3.1
	ClCH$_2$—	9.1	0.0	0.2	−0.2
	HOCH$_2$—	13.0	−1.4	0.0	−1.2
	CH$_2$=CH—	7.6	−1.8	−1.8	−3.5
	Ph—	13.0	−1.1	0.5	−1.0
	HC≡C—	−6.1	3.8	0.4	−0.2
	OHC—	9.0	1.2	1.2	6.0
	MeCO—	9.3	0.2	0.2	4.2
	RO$_2$C—	2.1	1.2	0.0	4.4
	N≡C—	−16.0	3.5	0.7	4.3
N	H$_2$N—	19.2	−12.4	1.3	−9.5
	Me$_2$N—	22.4	−15.7	0.8	−11.8
	AcNH—	11.1	−16.5	0.5	−9.6
	O$_2$N—	19.6	−5.3	0.8	6.0
O	HO—	26.9	−12.7	1.4	−7.3
	MeO—	30.2	−14.7	0.9	−8.1
	AcO—	23.0	−6.4	1.3	−2.3
Hal	F—	35.1	−14.3	0.9	−4.4
	Cl—	6.4	0.2	1.0	−2.0
	Br—	−5.4	3.3	2.2	−1.0
	I—	−32.3	9.9	2.6	−0.4
Si	Me$_3$Si—	13.4	4.4	−1.1	−1.1
P	Ph$_2$P—	8.7	5.1	−0.1	0.0
S	MeS—	9.9	−2.0	0.1	−3.7

Table 3.13 ^{13}C Chemical shifts of carbonyl carbons

R^1	R^2	δ_C	R^1	R^2	δ_C
Me—	—H	199.7	Me—	—OH	178.1
Et—	—H	206.0	Et—	—OH	180.4
Pr^i—	—H	204.0	Pr^i—	—OH	184.1
			Bu^t—	—OH	185.9
CH$_2$=CH—	—H	192.4	CH$_2$=CH—	—OH	171.7
Ph—	—H	192.0	Ph—	—OH	172.6
Me—	—Me	206.0	Me—	—OMe	170.7
Et—	—Me	207.6	Et—	—OMe	173.3
Pr^i—	—Me	211.8	Pr^i—	—OMe	175.7
Bu^t—	—Me	213.5	Bu^t—	—OMe	178.9
ClCH$_2$—	—Me	200.7	CH$_2$=CH—	—OMe	165.5
Cl$_2$CH—	—Me	193.6	Ph—	—OMe	166.8
Cl$_3$C—	—Me	186.3			
CH$_2$=CH—	—Me	197.2	—(CH$_2$)$_3$O—		177.9
Ph—	—Me	197.6	—(CH$_2$)$_4$O—		175.2
			Me—	—NH$_2$	172.7
—(CH$_2$)$_3$—		208.2	CH$_2$=CH—	—NH$_2$	168.3
—(CH$_2$)$_4$—		213.9	Ph—	—NH$_2$	169.7
—(CH$_2$)$_5$—		208.8	—(CH$_2$)$_3$NH—		179.4
—(CH$_2$)$_6$—		211.7	—(CH$_2$)$_4$NH—		173.0
cyclopent-2-enone		209.0	Me—	—OAc	167.3
			Ph—	—OAc	162.8
			Me—	—Cl	168.6
			CH$_2$=CH—	—Cl	165.6
			Ph—	—Cl	168.0
cyclohex-2-enone		198.0	Me—	—SiMe$_3$	247.6
			Ph—	—SiMe$_3$	237.9
			Me—	—SiPh$_3$	240.1

Table 3.14 ^1J ^{13}C-^1H Coupling constants

$^1J_{CH}$ CH$_4$ 125   CH$_2$=CH$_2$ 156   ⟨benzene⟩—H 159   HC≡CH 249

△ 161   ☐ 134   ⬠ 128   ⬡ 123

pyrrole (NH): 170, 182   furan (O): 167, 198   pyridine (N): 152, 163, 179

Estimation of $^1J_{CH}$ in alkanes

For $R^1R^2R^3C$—H, $$^1J_{CH} = 125 + \Sigma z_i \tag{3.19}$$

Table 3.15 Substituent constants z_i for Eq. 3.19

	Substituent R^i	z		Substituent R^i	z
H	H—	0	**N**	H$_2$N—	8
				Me$_2$N—	6
C	Me—	1			
	Bu^t—	−3	**O**	HO—	18
	ClCH$_2$—	3			
	HC≡C—	7	**Hal**	F—	24
	Ph—	1		Cl—	27
	OHC—	2		Br—	27
	MeCO—	−1		I—	26
	HO$_2$C—	6			
	NC—	11	**S**	MeSO—	13

Table 3.16 ^2J and ^3J ^{13}C-^1H Coupling constants

$^2J_{CH}$ 1-6 Hz   0-16 Hz   40-60 Hz

X=C or H at the low end of the range
X=Hal at the high end of the range

20-50 Hz   5-8 Hz   1-4 Hz

$^3J_{CH}$ 8 Hz   3 Hz   7-10 Hz

Table 3.17 ^{13}C-^{19}F Coupling constants in Hz

Structure	$^1J_{CF}$	Structure	$^1J_{CF}$	$^2J_{CF}$	$^3J_{CF}$	$^4J_{CF}$
CH_3F	158	$RCH_2CH_2CH_2F$	165	18	6	<2
CH_2F_2	237	$(RCH_2)_2CHF$	170	20		
CH_3F	274	HO_2CCH_2F	180			
CF_4	257	HO_2CCHF_2	247			
$MeCF_3$	281	HO_2CCF_3	283	45		
CCl_3F	337					
CCl_2F_2	325	⬡—F	245	21	8	3
$CClF_3$	299					
$MeC(=O)F$	353					
$(CF_3)_2O$	265	⬡—CF_3	272	32	4	1
$H_2C=CF_2$	287					

Table 3.18 ^{13}C-^{31}P Coupling constants in Hz

Structure	$^1J_{CP}$	$^2J_{CP}$	$^3J_{CP}$	$^4J_{CP}$
$(CH_3CH_2CH_2CH_2)_3P$	14	15	10	0
Me_4P^+	56			
$(CH_3CH_2CH_2CH_2)_3P=O$	66	5	13	0
$(CH_3CH_2CH_2CH_2O)_3P=O$		6	7	0
$CH_3CH_2CH_2CH_2PO(OBu)_2$	11	5	0	0
⬡—PPh_3	12	20	7	0

Table 3.19 ^{1}H Chemical shifts in methyl, methylene and methine groups

	Methyl protons	δ_H	*Methylene protons*	δ_H	*Methine protons*	δ_H
C	R—CH$_3$	0.9	R—CH$_2$—R	1.4	R—CHR$_2$	1.5
	C=C—C—CH$_3$	1.1	C=C—C—CH$_2$—R	1.7		
	O—C—CH$_3$	1.3	O—C—CH$_2$—R	1.9	O—C—CHR$_2$	2.0
	N—C—CH$_3$	1.1	N—C—CH$_2$—R	1.4		
	O$_2$N—C—CH$_3$	1.6	O$_2$N—C—CH$_2$—R	2.1		
	C=C—CH$_3$	1.6	C=C—CH$_2$—R	2.3		
	Ar—CH$_3$	2.3	Ar—CH$_2$—R	2.7	Ar—CHR$_2$	3.0
	O=CC=C—CH$_3$	2.0	O=CC=C—CH$_2$—R	2.4		
	O=CC(CH$_3$)=C	1.8	O=CC(CH$_2$—R)=C	2.4		
	C≡C—CH$_3$	1.8	C≡C—CH$_2$—R	2.2	C≡C—CHR$_2$	2.6
	RCO—CH$_3$	2.2	RCO—CH$_2$—R	2.4	RCO—CHR$_2$	2.7
	ArCO—CH$_3$	2.6	ArCO—CH$_2$—R	2.9	ArCO—CHR$_2$	3.3
	ROOC—CH$_3$	2.0	ROOC—CH$_2$—R	2.2	ROOC—CHR$_2$	2.5
	ArOOC—CH$_3$	2.4	ArOOC—CH$_2$—R	2.6		
	N—CO—CH$_3$	2.0	N—CO—CH$_2$—R	2.2	N—CO—CHR$_2$	2.4
	N≡C—CH$_3$	2.0	N≡C—CH$_2$—R	2.3	N≡C—CHR$_2$	2.7
N	N—CH$_3$	2.3	N—CH$_2$—R	2.5	N—CHR$_2$	2.8
	ArN—CH$_3$	3.0	ArN—CH$_2$	3.5		
	RCON—CH$_3$	2.9	RCON—CH$_2$—R	3.2	RCO—N—CHR$_2$	4.0
	N$^+$—CH$_3$	3.3	N$^+$—CH$_2$—R	3.3		
	O$_2$N—CH$_3$	4.3	O$_2$N—CH$_2$—R	4.4	O$_2$N—CHR$_2$	4.7
O	HO—CH	3.4	HO—CH$_2$—R	3.6	HO—CHR$_2$	3.9
	RO—CH$_3$	3.3	RO—CH$_2$—R	3.4	RO—CHR$_2$	3.7
	C=CO—CH$_3$	3.8	C=CO—CH$_2$—R	3.7		
	ArO—CH$_3$	3.8	ArO—CH$_2$—R	4.3	ArO—CHR$_2$	4.5
	RCOO—CH$_3$	3.7	RCOO—CH$_2$—R	4.1	RCOO—CHR$_2$	4.8
			(RO)$_2$CH$_2$	4.8	(RO)$_3$CH	5.2
Hal	F—CH$_3$	4.3	F—CH$_2$—R	4.1	F—CHR$_2$	3.7
	Cl—CH$_3$	3.1	Cl—CH$_2$—R	3.6	Cl—CHR$_2$	4.2
	Br—CH$_3$	2.7	Br—CH$_2$—R	3.5	Br—CHR$_2$	4.3
	I—CH$_3$	2.1	I—CH$_2$—R	3.2	I—CHR$_2$	4.3
S	RS—CH$_3$	2.1	S—CH$_2$—R	2.4	S—CHR$_2$	3.2
	RSO—CH$_3$	2.5	RSO—CH$_2$—R	2.7		
	RSO$_2$—CH$_3$	2.8	RSO$_2$—CH$_2$—R	2.9		
			(RS)$_2$CH$_2$	4.2		
P	R$_2$P—CH$_3$	1.4	R$_2$P—CH$_2$—R	1.6	R$_2$P—CHR$_2$	1.8
Si	R$_3$Si—CH$_3$	0.0	R$_3$Si—CH$_2$—R	0.5	R$_3$Si—CHR$_2$	1.2
Se	RSe—CH$_3$	2.0				

R = alkyl group. These values will usually be within ±0.2 p.p.m. unless electronic or anisotropic effects from other groups are strong. An obsolete scale used τ values; these are related to δ values by the equation $\tau = 10 - \delta$.

Estimation of ¹H chemical shifts in alkanes

For $R^1R^2R^3C-H$, $\qquad \delta_H = 1.50 + \Sigma z_i$ $\qquad\qquad$ (3.20)

Table 3.20 Substituent constants z for Eq. 3.20

R^i	z	R^i	z	R^i	z
H—	−0.3	HC≡C—	0.9	MeO—	1.5
Alkyl—	0.0	OHC—	1.2	PhO—	2.3
CH₂=CHCH₂—	0.2	MeCO—	1.2	AcO—	2.7
MeCOCH₂—	0.2	RO₂C—	0.8	Cl—	2.0
HOCH₂—	0.3	NC—	1.2	Br—	1.9
ClCH₂—	0.5	H₂N—	1.0	I—	1.4
CH₂=CH—	0.8	O₂N—	3.0	MeS—	1.0
Ph—	1.3	HO—	1.7	Me₃Si—	0.7

Table 3.21 ¹H Chemical shifts of -CH₂- and =CH- groups in some aliphatic cyclic compounds

at −100° H_{ax} 1.1
$\qquad\qquad$ H_{eq} 1.6

Axial protons generally come into resonance at higher field than their equatorial counterparts

Estimation of ¹H chemical shifts in alkenes

$$\delta_H = 5.25 + \Sigma z_{gem} + \Sigma z_{cis} + \Sigma z_{trans} \tag{3.21}$$

Table 3.22 Substituent constants z for Eq. 3.21

	Substituent R	z_{gem}	z_{cis}	z_{trans}
H	H—	0	0	0
C	Alkyl—	0.45	−0.22	−0.28
	[a]Ring-alkyl—	0.69	−0.25	−0.28
	N≡CCH₂— or RCOCH₂—	0.69	−0.08	−0.06
	ArCH₂—	1.05	−0.29	−0.32
	R₂NCH₂—	0.58	−0.10	−0.08
	ROCH₂—	0.64	−0.10	−0.02
	HalCH₂—	0.70	0.11	−0.04
	RSCH₂—	0.71	−0.13	−0.22
	Isolated RCH=CH—	1.00	−0.09	−0.23
	[b]Conjugated CH=CH—	1.24	0.02	−0.05
	Ar—	1.38	0.36	−0.07
	OHC—	1.02	0.95	1.17
	Isolated RCO—	1.10	1.12	0.87
	[b]Conjugated RCO—	1.06	0.91	0.74
	Isolated HO₂C—	0.97	1.41	0.71
	[b]Conjugated HO₂C—	0.80	0.98	0.32
	Isolated RO₂C—	0.80	1.18	0.55
	[b]Conjugated RO₂C—	0.78	1.01	0.46
	R₂NCO—	1.37	0.98	0.46
	ClCO—	1.11	1.46	1.01
	RC≡C—	0.47	0.38	0.12
	N≡C—	0.27	0.75	0.55
N	(Alkyl)HN— or (Alkyl)₂N—	0.80	−1.26	−1.21
	[b](Conjugated alkyl or aryl)₂N—	1.17	−0.53	−0.99
	AcNH—	2.08	−0.57	−0.72
	O₂N—	1.87	1.30	0.62
O	AlkylO—	1.22	−1.07	−1.21
	[b]Conjugated alkyl or arylO—	1.21	−0.60	−1.00
	AcO—	2.11	−0.35	−0.64
Hal	F—	1.54	−0.40	−1.02
	Cl—	1.08	0.18	0.13
	Br—	1.07	0.45	0.55
	I—	1.14	0.81	0.88
Si	R₃Si—	0.90	0.90	0.60
S	RS—	1.11	−0.29	−0.13
	RSO—	1.27	0.67	0.41
	RSO₂—	1.55	1.16	0.93

[a]Use the 'ring-alkyl' values when the double bond and the alkyl group are part of a five- or six-membered ring. [b]Use the 'conjugated' values when either the substituent or the double bond is further conjugated.

Table 3.23 ^{1}H Chemical shifts of protons attached to multiple bonds

Structure	δ_H	Structure	δ_H
RCHO	9.4-10.0	R$_2$C=CHR	4.5-6.0
ArCHO	9.7-10.5	R$_2$C=CH—COR	5.8-6.7
ROCHO	8.0-8.2	RHC=CR—COR	6.5-8.0
R$_2$NCHO	8.0-8.2	RHC=CR—OR	4.0-5.0
RC≡CH	1.8-3.1	R$_2$C=CH—OR	6.0-8.1
R$_2$C=C=CHR	4.0-5.0	RHC=CR—NR$_2$	3.7-5.0
ArH	6.0-9.0	R$_2$C=CH—NR$_2$	5.7-8.0

Table 3.24 ^{1}H Chemical shifts (largely attached to double bonds) in some unsaturated cyclic systems (for simple cycloalkenes see Table 3.21)

Estimation of ^{1}H chemical shifts in substituted benzenes

$$\delta_H = 7.27 + \Sigma z_i \qquad (3.22)$$

Table 3.25 Substituent constants z for Eq. 3.22

	Substituent R	z_{ortho}	z_{meta}	z_{para}
H	H—	0	0	0
C	Me—	−0.20	−0.12	−0.22
	Et—	−0.14	−0.06	−0.17
	Pr^i—	−0.13	−0.08	−0.18
	Bu^t—	−0.02	−0.08	−0.21
	H_2NCH_2— or HOCH$_2$—	−0.07	−0.07	−0.07
	ClCH$_2$—	0.00	0.00	0.00
	F$_3$C—	0.32	0.14	0.20
	Cl$_3$C—	0.64	0.13	0.10
	CH$_2$=CH—	0.06	−0.03	−0.10
	Ph—	0.37	0.20	0.10
	OHC—	0.56	0.22	0.29
	MeCO—	0.62	0.14	0.21
	H$_2$NCO—	0.61	0.10	0.17
	HO$_2$C—	0.85	0.18	0.27
	MeO$_2$C—	0.71	0.10	0.21
	ClCO—	0.84	0.22	0.36
	HC≡C—	0.15	0.02	−0.01
	N≡C—	0.36	0.18	0.28
N	H$_2$N—	−0.75	−0.25	−0.65
	Me$_2$N—	−0.66	−0.18	−0.67
	AcNH—	0.12	−0.07	−0.28
	O$_2$N—	0.95	0.26	0.38
O	HO—	−0.56	−0.12	−0.45
	MeO—	−0.48	−0.09	−0.44
	AcO—	−0.25	0.03	−0.13
Hal	F—	−0.26	0.00	−0.04
	Cl—	0.03	−0.02	−0.09
	Br—	0.18	−0.08	−0.04
	I—	0.39	−0.21	0.00
Si	Me$_3$Si—	0.22	−0.02	−0.02
	(MeO)$_2$P(=O)—	0.48	0.16	0.24
S	MeS—	0.37	0.20	0.10

These parameters are simply those measured on the corresponding monosubstituted benzene ring; they are not accurately taken over to polysubstituted benzenes, but the estimation of chemical shift is usually fairly good. Errors are particularly likely to occur when substituents *ortho* to one another interfere with conjugation to the ring.

Table 3.26 ^{1}H and ^{13}C Chemical shifts in the common deuterated solvents

Solvent	Deuterated solvent				Undeuterated solvent
	$\delta_H{}^a$	Multi-plicity^b	δ_C	Multi-plicity^b	δ_C
Acetic acid	2.05				21.1
	11.5^c				178.1
Acetone	2.05	quintet	29.8	septet	30.5
			205.7		205.4
Acetonitrile	1.95	quintet	1.3	septet	1.7
			118.2	m	118.2
Benzene	7.27		128.0	triplet	128.5
t-Butanol	1.28^d				
Carbon disulfide					192.8
Carbon tetrachloride					96.1
Chloroform	7.25		77.0	triplet	77.2
Cyclohexane	1.40	triplet	26.3	quintet	27.6
1,2-Dichloroethane	3.72	br			
Dichloromethane (methylene chloride)	5.35	triplet	53.1	quintet	54.0
Dimethylformamide (DMF)	2.75	quintet			
Dimethylsulfoxide (DMSO)	2.5	quintet	39.7	septet	40.6
water in DMSO	3.3^c				
Dioxan	3.55	triplet	66.5	quintet	67.6
Hexamethylphosphoricamide (HMPA)	2.60	doublet^e quintet	35.8	septet	36.9
Methanol	3.35	quintet	49.0	septet	49.9
	4.8^c				
Nitromethane	4.33	quintet	60.5	septet	57.1
Pyridine	7.0		123.4	triplet	123.9
	7.35		135.3	triplet	135.9
	8.5		149.8	triplet	150.3
Tetrahydrofuran (THF)	1.73	br	25.2	quintet	26.5
	3.58	br	67.4	quintet	68.4
Toluene	2.3	quintet			
	7.2				
Trifluoroacetic acid (TFA)	11.3^c				115.7^f
					163.8^g
Water	4.7^c				

^{a}Residual protons in the deuterated solvent.
^{b}A singlet unless otherwise stated.
^{c}Variable, depends upon the solvent and its concentration.
^{d}(CH$_3$)$_3$COD is usually used, not the fully deuterated solvent.
^{e}Coupled to P, $J = 9$ Hz.
^{f}Quartet from coupling to F, $J = 294$ Hz.
^{g}Quartet from coupling to F, $J = 46$ Hz.

Table 3.27 ^{1}H Chemical shifts of protons attached to elements other than carbon^a

	Structure	δ_H		Structure	δ_H
NH	RNH$_2$ and R$_2$NH	0.5-4.5	**OH**	Monomeric H$_2$O	~1.5
	ArNH$_2$ and ArNHR	3-6		Suspended H$_2$O	~4.7
	RCONH$_2$ and RCONHR	5-12		ROH	0.5-4.5
	Pyrrole NH	7-12		ArOH	4.5-10
SiH	Me$_3$SiH	4.0		RCO$_2$H	9-15
	Ar$_3$SiH	~5.5		R$_2$C=NOH	9-12
	(MeO)$_3$SiH	4.11			
	Cl$_2$SiH$_2$	5.40			7-16
	MeOSiH$_3$	4.52			
SnH	R$_3$SnH	~5.3	**SH**	RSH	1-2
PH	(RO)$_2$P(=O)H	~6.8^b		ArSH	3-4

^{a}These values (except for SiH and SnH) are sensitive to temperature, solvent and concentration; the stronger the hydrogen bond, the lower field the chemical shift. ^{b}Doublet, $^1J_{PH}$ 140 Hz.

Table 3.28 Geminal ($^2J_{HH}$) coupling constants in Hz

R^1, R^2	$^2J_{HH}$		$^2J_{HH}$
H— —H	−12.4		
R— —R	−8...−18		−21.5
—(CH$_2$)$_2$—	−3...−9		
—(CH$_2$)$_3$—	−11...−17		
—(CH$_2$)$_4$—	−8...−18		
—(CH$_2$)$_5$—	−11...−14		
H— —Ph	−14.3		−3...+3
H— —OH	−10.8		
H— —Cl	−10.8		
—O(CH$_2$)$_2$O—	~0		
—O(CH$_2$)$_3$O—	−5...−6		−8...−10
H— —CN	−16.2		
H— —COMe	−14.9		

Table 3.29 Vicinal ($^3J_{HH}$) coupling constants in Hz in some aliphatic compounds

Open-chain compounds			Cyclic compounds			
Structure	$^3J_{HH}$ range	Typical value	Structure	Geometry	Ring size	$^3J_{HH}$ range
CH₃—CH₂—	6-8	7		cis	3	7-13
				trans	3	4.0-9.5
CH₃-CH<	5-7	6		cis	4	4.0-12.0
				trans	4	2.0-10.0
—CH₂—CH₂—	5-8	7		cis	5	5.0-10.0
				trans	5	5.0-10.0
>CH-CH<	0-8	7		cis	6	2.0-6.0
				trans	6	8.0-13.0[b]
>=CH—CH<	4-11	6			3	1.8[c]
					4	−0.8[c]
>=CH—CH=<	6-13	11[a]			5	0.5[c]
					6	1.5[c]
>CH·CHO	0-3	2			7	3.7[c]
					8	5.3[c]
>=CH—CHO	5-8	7			3	0.5-2
					4	2.5-4.0
(cis alkene)	0-12	8			5	5-7
					6	8.5-10.5
					7	9-12.5
					8	10-13
					1-2x	3-4
(trans alkene)	12-18	15			1-2n	0-2
					2x-3x	9-10
					2n-3n	6-7
					2x-3n	2-5
					1-7	0-3

[a] Found in dienes adopting the s-*trans* conformation.
[b] J_{aa}=8-13, J_{ee} = 2-5; note that J_{ee} is usually 1 Hz smaller than J_{ae}.
[c] Value for the unsubstituted cycloalkene.

Table 3.30 Vicinal ($^3J_{HH}$) coupling constants in Hz in some heterocyclic and aromatic compounds

Table 3.31 Pascal's triangle giving the relative intensities of first-order multiplets for coupling to n equivalent nuclei of spin $I = 1/2$

n	Relative intensity
0	1
1	1 1
2	1 2 1
3	1 3 3 1
4	1 4 6 4 1
5	1 5 10 10 5 1
6	1 6 15 20 15 6 1
7	1 7 21 35 35 21 7 1
8	1 8 28 56 70 56 28 8 1

Table 3.32 Long-range ($^4J_{HH}$ and $^5J_{HH}$) coupling constants in Hz

Structure	$^4J_{HH}$	Structure	$^5J_{HH}$
—CH=C—CH	0-3	CH-C=C—CH	0-2
—CH=C=CH—	4-6	—CH=C=C—CH	2-3
H—C≡C—CH	1-3	CH-C≡C—CH	1-3
(meta aromatic H, H)	1-3	(para-quinonoid H···H)	8-10
(benzyl CH₂)	0.6-0.9 (small, because not W)	(para aromatic H···H)	0-1
(cyclohexane H, H)	1-2		
(bicyclic W arrangement)	7-8		
H^{7a}, H^{6x}, H^{2x}, H^{2n}	7a-2n 3-4 2x-6x 1-2	H^1, H^4 (norbornane)	1-1.5
CH₃ (decalin)	Signal perceptibly broadened by ^4J coupling		

Table 3.33 Eu(dpm)$_3$-induced shifts of protons in some common environments[a]

Functional group	Shift p.p.m./mol of Eu(dpm)$_3$ per mol of substrate	Functional group	Shift p.p.m./mol of Eu(dpm)$_3$ per mol of substrate
RCH₂NH₂	~150	RCH₂CHO	11
RCH₂OH	~100	RCH₂OCH₂R	10
RCH₂NH₂	30-40	RCH₂CO₂CH₃	7
RCH₂OH	20-25	RCH₂CO₂CH₃	6.5
RCH₂COR	10-17	RCH₂CN	3-7
RCH₂CHO	19		

[a]The shifts refer to the protons in italics.

Table 3.34 ^{1}H–^{19}F Coupling constants in Hz

Structure		J	Structure		J
$^2J_{HF}$	H–F (on tetrahedral C)	45-52	$^4J_{HF}$	$>$CH–C–CF$<$	0-9^b
	H–F (cyclopropane)	60-65		H–C=C–CF (allylic)	2-4
	H—C=C—F (vinylic)	72-90		H—C=C—CF	0-6
$^3J_{HF}$	CH$_3$—CF$<$	20-24	$^{3\text{-}6}J_{HF}$		
	$>$CH–CF$<$	0-45^a		H aromatic (C–F)	ortho 6-11 meta 3-9 para 0-4
	H—C=C—F (cis)	3-20			
	H—C=C—F (trans)	12-53		CH$_3$ aromatic (C–F)	ortho 2.5 meta 1.5 para 0

^{a}0-12 when gauche and 10-45 when antiperiplanar.
^{b}The higher end of the range ($\geq$3.5) when the atoms are held in a W conformation.

Table 3.35 ^{1}H–^{31}P Coupling constants in Hz^a

Type of coupling	Class of compound		
	Phosphines	Phosphonium salts	Phosphine oxides
$^1J_{PH}$	(150) 185-220 (250)	400-900	200-750
$^2J_{PCH}$	(−5) 0-15 (27) and 46^b	(0) 10-18 & 30^b	5-25 and 40^b
$^3J_{PCCH}$	(10) 13-17 (20)	(0) 10-20 57	14-30
$^3J_{PC=CH}$	trans (5) 12-41 cis 6-20^c	trans 28-50 (80) cis 10-20 (35)^c	
		Phosphites	Phosphates
$^3J_{POCH}$		(0) 5-14 (20)	(0) 5-20 (30)
	All compounds		
$^4J_{PH}$	0-3 (5)^d		

^{a}The coupling constants are often strongly dependent upon the groups attached to phosphorus, and therefore values outside the quoted range may occasionally be observed; values in parentheses are 'extreme' values so far reported.
^{b}Values observed in the PCH=C system.
^{c}trans Coupling constants are usually about twice the cis coupling constants.
^{d}In the system P—C=C=CH.

Table 3.36 Some representative ^{11}B chemical shifts in p.p.m. relative to $Et_2O.BF_3$ (negative numbers are upfield and positive numbers downfield)

Structure	δ	Structure	δ	Structure	δ
Me_3B	87	$Me_2B(OMe)$	53	$Me_3B.NMe_3$	0.1
Me_2BF	60	$MeB(OMe)_2$	30	$Me_3B.PMe_3$	12
$MeBF_2$	8	$B(OMe)_3$	18	$H_3B.NMe_3$	−8
BF_3	10	$MeB(NMe_2)_2$	34	$H_3B.SMe_2$	−20
Ph_3B	68	BCl_3	47	BF_4^-	−2
$(CH_2=CH)_3B$	56	BBr_3	39		
B_2H_6	17				

For more about ^{11}B NMR, see S. Hermanek, *Chem. Rev.*, 1992, 92, 325; H. Nöth and B. Wrackmeyer, *NMR Basic Principles and Progress*, 1978, **14**, 1.

Table 3.37 Some representative one-bond ^{11}B coupling constants in Hz

Structure		J	Structure		J
Me_2BF	$^1J_{BF}$	119	Me_3B	$^1J_{CB}$	47
$MeBF_2$	$^1J_{BF}$	76	BH_4^-	$^1J_{BH}$	80
BF_3	$^1J_{BF}$	15	Me_4B^-	$^1J_{CB}$	22
BF_4^-	$^1J_{BF}$	1-5[a]			

[a]Solvent and temperature dependent.

Table 3.38 Approximate ^{15}N chemical shifts in p.p.m. relative to $MeNO_2$ (negative numbers are upfield and positive numbers downfield)

Structure	δ	Structure	δ	Structure	δ
R_3N	−350	$RN=C=NR$	−250	Pyridines	50 to −50
R_4N^+	−350	$RC≡N$	−150	$R_2C=NOH$	0
$RNHNH_2$	−350	Pyrroles	−100 to −250	$R_2C=NR$	0 to −50
$RNCO$	−350	$RCNO$	−180	$RN=NR$	200
RN_3	−350 −190 −160	$(RCO)_2NR$	−180	$R_2N—N=O$	200
$R_3N^+—O^-$	−260	$R\overset{O^-}{\underset{R\;+}{\overset{|}{N}}}\diagup R$	−100	$R—N=O$	500
$RCONR_2$	−330				

Table 3.39 Some representative ^{19}F chemical shifts in p.p.m. relative to $CFCl_3$ (negative numbers are upfield and positive numbers downfield)

Structure	δ	Structure	δ	Structure	δ
MeF	−272	$(CF_3)_3N$	−56	HF	40
CF_2H_2	−144	PhF	−116	F_2	429
CHF_3	−79	C_6F_6	−163	BF_3	−131
CF_4	−63	$PhCF_3$	−64	$BF_3.OEt_2$	−153
EtF	−213	$PhCH_2F$	−207	SiF_4	−163
c-C_6F_{12}	−133	$BrCF_3$	7	Et_2SiF_2	−143
$CH_2=CHF$	−114	CF_3CO_2H	−78	PF_3	−34
$CH_2=CF_2$	−81	CF_3CO_2Me	−74	PF_5	−72
$CF_2=CF_2$	−135	$(CF_3)_2CO$	−85	POF_3	−91
cis-CHF=CHF	−165	F—C≡N	−156	SF_6	57
$trans$-CHF=CHF	−183	F—C≡C—F	−95	SO_2F_2	33
$CF_3CF_2CF_2CF_3$	−135	NF_3	145	$PhSO_2F$	65

For more about ^{19}F NMR, see *Handbook of Basic Tables for Chemical Analysis*, Ed. T. J. Bruno and P. D. N. Svoronos, CRC Press, Boca Raton, 2nd Ed., 2003; J. W. Emsley and L. Phillips, *Progress in NMR Spectroscopy*, 1971, 7, 1; J. W. Emsley, L. Phillips and V. Wray, *Progress in NMR Spectroscopy*, 1976, **10**, 83.

Table 3.40 Some representative ^{19}F to ^{19}F coupling constants in Hz (for coupling constants from ^{19}F to ^{13}C, ^{1}H and ^{11}B, see Tables 3.17, 3.34 and 3.37)

Structure		$^2J_{FF}$	Structure	$^3J_{FF}$
$(CH_2)_n CF_2$	n = 3	150	CF_3CF_3	3.5
	n = 4	200	CF_3CHF_2	3
	n = 5	240	CF_3CH_2F	16
	n = 6	228	FCH_2CH_2F	11
(F₃C, F, F, F structure)		270	(F F structure)	134
	R,R = H,H	36	(F F structure)	19
	R,R = H,F	87	$ClCF_2CF_2C(=O)F$	5
(R F / R F alkene)	R,R = Br,F	75	$ClCF_2CF_2C(=O)F$	8
	R,R = Br,Cl	30	$CHF_2CHFCHF_2$	13
	R,R = Ph,H	33	CF_3CF_2CHFMe	15
	R,R = n-C_6H_{13},H	50	$CF_3CF_2CF_2CO_2H$	<1
	R,R = CF_3CH_2O,F	102	CF_3CF_2CHFMe	<1

$^4J_{FF}$ and $^5J_{FF}$ are often much larger than $^3J_{FF}$, especially when F is attached to trigonal carbon. They are very dependent upon the structure, including the spatial proximity of the F atoms.

Table 3.41 Some representative ^{29}Si chemical shifts in p.p.m. relative to Me$_4$Si (negative numbers are upfield and positive numbers downfield)

Structure	δ	Structure	δ
Me$_4$Si	0	Cl$_3$SiH	−10
t-PhCH=CHSiMe$_3$	−7	Me$_3$SiSiMe$_3$	−20
Me$_3$SiH	−17	PhMe$_2$SiSiMe$_3$	−19
Me$_2$SiH$_2$	−37	PhMe$_2$SiSiMe$_3$	−22
MeSiH$_3$	−65	Me$_3$SiMe$_2$SiSiMe$_3$	−16
Me$_3$SiF	31	Me$_3$SiMe$_2$SiSiMe$_3$	−49
Me$_2$SiF$_2$	9	c-(Me$_2$Si)$_6$	−42
Me$_3$SiCl	30	Ph$_3$SiH	−21
Me$_2$SiCl$_2$	32	Ph$_2$SiH$_2$	−33
MeSiCl$_3$	12	PhSiH$_3$	−60
Me$_3$SiBr	26	PhMe$_2$SiH	−17
Me$_3$SiI	9	Ph$_2$SiCl$_2$	6
Me$_3$SiOMe	17	Ph$_3$SiLi	−9
(EtO)$_3$SiH	−59	PhMe$_2$SiLi	−29
(MeO)$_4$Si	−78	(PhMe$_2$Si)$_2$CuLi$_2$	−24
Me$_3$SiOClO$_3$	47	AcNHSiMe$_3$	6
(Me$_3$Si)$_2$O	7	Me(C=NH)OSiMe$_3$	18
c-(Me$_2$SiO)$_4$	−20	AcN(SiMe$_3$)$_2$	6

For more about ^{29}Si NMR, see E. A. Williams, NMR Spectroscopy of Organosilicon Compounds, Ch. 8 in *The Chemistry of Organosilicon Compounds*, Ed. S. Patai and Z. Rappoport, Wiley, New York, 1989; M. A. Brook, *Silicon in Organic, Organometallic, and Polymer Chemistry*, Wiley, New York, 2000.

Table 3.42 Some representative coupling constants from ^{29}Si to $^1H^a$, ^{13}C, ^{19}F, ^{29}Si and ^{31}P in Hz

Structure	$^1J_{SiH}$	Structure	$^1J_{SiH}$	Structure	$^2J_{SiH}$
SiH_4	202	$MeOSiH_3$	216	R_3SiCH_2R	~10
Me_3SiH	184	$(MeO)_3SiH$	298		$^3J_{SiH}$
Cl_2SiH_2	288			$R_3SiCH_2CH_2R$	0-5
	$^1J_{SiF}$		$^1J_{SiF}$		
CCl_3SiF_3	264	$ClCH_2SiF_3$	267		
	$^1J_{CSi}$		$^1J_{SiSi}$		$^1J_{PSi}$
$(CH_3)_4Si$	50	$Ph_3SiSiMe_3$	87	$(Me_3Si)_3P$	27
		$(Me_3Si)_4Si$	53		

[a] $^1J_{SiH}$ is seen as two weak bands, one on each side of the main proton signal, similar to the ^{13}C sidebands in ^1H spectra but more intense (the natural abundance of ^{29}Si is 4.7%).

Table 3.43 Some representative $^{31}P(III)$ chemical shifts in p.p.m. relative to 85% H_3PO_4 (negative numbers are upfield and positive numbers downfield)

Structure	δ	Structure	δ	Structure	δ
Me_3P	−62	Me_2PH	−99	Me_2PF	186
Et_3P	−20	$MePH_2$	−164	$MePF_2$	245
$n\text{-}Pr_3P$	−33			$(RO)_3P$	125-145
$i\text{-}Pr_3P$	19			$PHal_3$	120-225
$t\text{-}Bu_3P$	63				

For more about ^{31}P NMR, see *CRC Handbook of P-31 NMR Data*, Ed. J. C. Tebby, CRC Press, Boca Raton, 1991.

Table 3.44 Some representative $^{31}P(V)$ chemical shifts in p.p.m. relative to 85% H_3PO_4 (negative numbers are upfield and positive numbers downfield)

Structure	δ	Structure	δ	Structure	δ
$Me_3P{=}O$	36	$Hal_3P{=}O$	−80-+5	PCl_5	−80
$Et_3P{=}O$	48	$(RO)_3P{=}O$	−20-0	PCl_4^+	86
$Et_3P{=}S$	55	$(RO)_3P{=}S$	60-75	PCl_6^-	−295
Me_4P^+	24	$Ar_3P{=}CR_2$	5-25		

Table 3.45 Some representative coupling constants from ^{31}P to ^{19}F in Hz

Structure	$^1J_{PF}$	Structure	$^2J_{PF}$	Structure	$^3J_{PF}$
Alkyl$_2$PF	821–1450	R$_2$PCFR$_2$	40–149	CHF$_2$CH$_2$PH$_2$	8
Ph$_2$PF	905	CF$_3$PF$_2$	87	CHF$_2$CH$_2$PCl$_2$	13
Me$_2$P(=O)F	980	FCH$_2$CF$_2$PCl$_2$	99		
Me$_3$PF$_2$	552				
Ph$_3$PF$_2$	660				
PF$_6^-$	706				

Table 3.46 A checklist of techniques and conditions, typical for an organic compound of MW ~300 using a 500 MHz instrument (less material may be used, with a corresponding increase in acquisition times)

Name		Description	Typical sample size	Typical run time
APT	1D ^{13}C	Plots ^{13}C signals for C & CH$_2$ on one side of a horizontal line, and CH and CH$_3$ on the other	30 mg	15 min
DEPT	1D ^{13}C	Plots ^{13}C signals separately for C, CH, CH$_2$ and CH$_3$	30 mg	15 min
1D TOCSY (HOHAHA)	1D ^{1}H-^{1}H	Plots ^{1}H signals separately for each ^{1}H-to-^{1}H spin system	10 mg	2 min
COSY	2D ^{1}H-^{1}H	Cross-peaks identify ^{1}H-to-^{1}H coupling	10 mg	20 min
NOESY and ROESY	2D ^{1}H-^{1}H	Cross-peaks identify ^{1}H-to-^{1}H proximity in space; NOESY for low MW compounds, ROESY for high MW compounds	10 mg	8 h[a]
2D TOCSY	2D ^{1}H-^{1}H	Cross-peaks identify members of the same ^{1}H-to-^{1}H spin system	10 mg	2 h
HMQC	2D ^{1}H-^{13}C	Cross-peaks identify ^{1}H-to-^{13}C one-bond connections	10 mg	40 min
HMBC	2D ^{1}H-^{13}C	Cross-peaks identify ^{1}H-to-^{13}C two- and three-bond connections	10 mg	2 h
HSQC-HECADE	2D $^{2\&3}J_{CH}$	Cross-peaks measure $^2J_{CH}$ and $^3J_{CH}$ values	30 mg	8 h[a]
INADEQUATE	2D ^{13}C-^{13}C	Cross-peaks identify ^{13}C-to-^{13}C one-bond connections	500 mg	48 h[a,b]
3D TOCSY	3D ^{1}H-^{1}H	Cross-peaks identify members of the same ^{1}H-to-^{1}H spin system, with each spin system in a different plane	30 mg	24 h[a]

[a]Acquisition time is much reduced with a cryoprobe. [b]To obtain a high-quality spectrum, a cryoprobe is nearly essential.

4. Mass spectra

4.1 Introduction

A mass spectrometer is a device for producing and weighing ions from a compound for which we wish to obtain molecular weight and structural information. All mass spectrometers use three basic steps: molecules M are taken into the gas phase; ions, such as the cations $M^{\cdot+}$, MH^+ or MNa^+, are produced from them (unless the molecules are already charged); and the ions are separated according to their mass-to-charge ratios (m/ze). Multiply charged ions are rarely produced except when electrospray mass spectrometry (ESI) or laser desorption (LD) is used to produce ions. Therefore, for other ionisation techniques, z can normally be taken as one; and since e is a constant (the charge on one electron), m/z gives the mass of the ion. Some of the devices that are used to produce gas-phase ions put enough vibrational energy into the ions to cause them to fragment in various ways to produce new ions with the loss of neutral fragments. Through this fragmentation, structural information can be obtained.

The spectrometer is designed to detect only the charged components (e.g. MH^+, and its associated fragments A^+, B^+, C^+, etc.), because only they are accelerated or deflected by the electromagnetic or electrostatic fields used in the analytical system. When the array of ions has been separated and recorded, the output is known as a mass spectrum. It is a record of the abundance of each ion reaching the detector (plotted vertically) against its m/z value (plotted horizontally). The mass spectrum is a result of a series of competing and consecutive unimolecular reactions, and what it looks like is determined by the chemical reactivity of the isolated sample molecules in the gas phase. It is not a spectroscopic method based on electromagnetic radiation, but since it complements information provided to the organic chemist by the UV, IR and NMR spectra, it is conveniently considered alongside them. Mass spectrometry is the most sensitive of all these methods. It can routinely be carried out with a few micrograms of sample, and in favourable cases even with picograms (10^{-12} g), making it especially important in solving problems where only a very small sample is available.

There is a variety of instruments available for taking a mass spectrum—they differ in the methods by which the molecules are taken into the gas phase, they differ by the way in which the molecules are induced to produce ions, and they differ by the method used to analyse the ions. Perhaps most significantly, as far as this book is concerned, they also differ in the degree to which they induce fragmentation. In this chapter, we shall see how several of these instruments work, and provide molecular weight and structural information. One of them, electron impact, will be described in more detail, because it is the everyday workhorse in structure determination of small and volatile molecules. Most of the others, with more specialised uses, will be described only in outline. The methods used to produce ions in the gas phase are covered first, and a consideration of the ways in which the ions may then be separated according to their mass-to-charge ratios (*m/ze*) follows. Finally, examples of mass spectra are given, through which the common combinations for ion production and analysis (chosen according to the type of compound that is being analysed) are illustrated.

4.2 Ion production from readily volatile molecules

4.2.1 Electron impact (EI)

This method of ion production is used in the analysis of relatively volatile organic molecules. These typically have molecular weights up to a maximum of around 400 Daltons, but, in cases like glucose with its numerous polar hydroxyl groups the volatility is too low, even though it has a much lower molecular weight. The sample is simply heated to evaporate it into the ionisation chamber (Fig. 4.1), which is kept at a very low pressure, typically $\leq 10^{-4}$ Nm^{-2} ($\leq 10^{-6}$ mmHg), to avoid collisions between molecules. Alternatively, and more commonly, the sample is placed on the ceramic tip of a long probe which can be inserted into the ion source. When the low pressure has stabilised, the tip is heated to 200-300 °C to drive the molecules off the surface into the ionisation chamber. Most organic molecules are stable at these temperatures provided that they do not collide with other molecules.

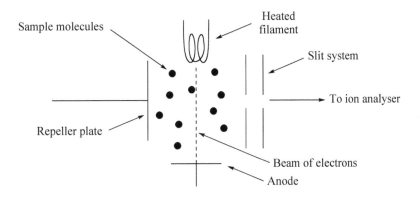

Sample molecules

Heated filament

Slit system

To ion analyser

Repeller plate

Beam of electrons

Anode

Fig. 4.1

At the same time, electrons are driven off a heated filament in the ionisation chamber by attraction to an anode, usually through a potential difference of about 70 eV (1 eV ≈ 23 kcal mol^{-1} ≈ 96 kJ mol^{-1}). A 70 eV electron has enough energy to expel an electron from a molecule with which it collides; this expulsion of an electron corresponds to the ionisation potential (IP) of the molecule, which typically requires around 7-10eV. Through this process, a radical-cation (M$^{\cdot+}$) is formed.

$$M + e \rightarrow M^{\cdot+} + 2e$$

A repeller plate at one end of the ionisation chamber is positively charged, and therefore repels the positively charged radical-cations through a slit system into the mass spectrometer, where they, and fragments derived from them, will be separated according to their *m/z* values.

Electron capture to give a radical-anion does not occur to a significant extent, because the bombarding electrons have too much translational energy to be captured. The 70 eV electron does not deposit all its energy into a molecule with which it interacts. In addition to the 7-10 eV required for ionisation of the sample, a further 0-6 eV is typically transferred as internal energy into the resulting ion. Since the strongest single bonds in organic molecules have strengths of about 4 eV, and many are much weaker, this is enough internal energy to lead to extensive fragmentation in most EI spectra.

4.2.2 Chemical Ionisation (CI)

In the numerous applications of mass spectrometry, frequently the single most important requirement is to obtain a molecular weight, and hence, if the mass measurement is sufficiently precise, a molecular formula. In such cases, the deposition of energy into the molecule should be smaller than in EI, and the form of the molecular ion that is generated should be intrinsically stable. Chemical ionisation (CI) fulfils both these criteria, and is also applicable to molecules that can be transferred to the gas phase by the simple expedient of heating.

In CI, a reagent gas, usually methane or more commonly ammonia, is passed into the ion chamber at a pressure of about 10^2 Nm^{-2}. This gas is ionised by using electrons, produced from a hot filament as before, but with energies up to 300 eV, giving rise (in the case of methane gas) to $CH_4^{\bullet+}$. At the operating pressures of CI sources, this ion collides with its neutral counterparts, which are present in much higher concentration than the sample molecules. The main bimolecular reaction that occurs is:

$$CH_4^{\bullet+} + CH_4 \rightarrow CH_5^+ + CH_3^{\bullet}$$

If sample molecules are volatilised into this mixture of ions, the CH_5^+ ion, which is essentially a proton 'solvated' by methane, acts as a strong acid and protonates the sample:

$$M + CH_5^+ \rightarrow MH^+ + CH_4$$

Thus, in positive-ion CI spectra, molecular weight information is obtained from protonation of sample molecules, and the observed m/z value is one unit higher than the molecular weight. When ammonia is used to generate CI spectra, the reagent ion which protonates the sample is NH_4^+. Because NH_4^+ binds a proton much more strongly than CH_5^+, much less energy is released in the transfer of a proton to M from the former than the latter. Some molecules M are not basic enough to be protonated by NH_4^+, so in these cases CH_5^+ must be employed instead. Conversely, where proton transfer can be effected from NH_4^+, the internal energy of MH^+, and hence the degree of fragmentation, is much less, and the informative MH^+ is more abundant in the mass spectrum. Thus, ammonia CI producing MH^+ (and/or MNH_4^+) is an excellent technique for the determination of molecular weights of volatile and functionalised molecules, with little fragmentation occurring. In any CI mass spectrum, the ions have an even number of electrons, and are for that reason, as well as from having a lower internal energy, less susceptible to fragmentation than in an EI mass spectrum.

Although EI does not produce satisfactory negative-ion spectra, negative-ion CI works well for molecules with electron-accepting properties, such as trifluoroacetates, quinones and nitro compounds. This is because the collisions occurring in a CI source reduce the large initial kinetic energies of the bombarding electrons to lower values at which they can be captured to give a radical-anion. Alternatively, a reagent ion such as CH_3O^- may be generated in the CI source, and this can act as a Brønsted base, abstracting a proton from the sample molecule:

$$M + CH_3O^- \rightarrow (M - H)^- + CH_3OH$$

A schematic diagram of a CI source is given in Fig. 4.2. The sample may be introduced directly by lowering the upper probe, which can be a gas line or can incorporate a heated ceramic tip; or indirectly from the output of a GC instrument by raising the lower probe. The ion chamber has holes in it to allow an electron beam to be directed through it. The whole of the ion chamber can be withdrawn from the centre of the vacuum chamber, the reagent gas supply can be cut off and the vacuum improved, allowing the same instrument to be used also for EI spectra. In either CI or EI mode, the path to the analyser will be pumped to the usual low pressure to minimise further collisions.

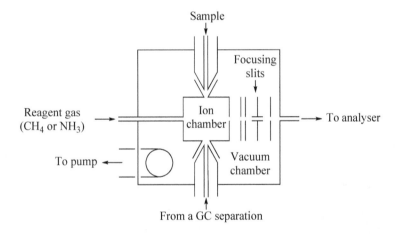

Fig. 4.2

4.3 Ion production from poorly volatile molecules

Several techniques, which are fundamentally different from EI and CI, have been developed to surmount the twin problems of getting involatile samples into the gas phase and in a charged state. These are: fast ion bombardment (FIB, also commonly called liquid secondary ion MS, or LSIMS), laser desorption (LD), and electrospray ionisation (ESI). LD and ESI are spectacularly successful in allowing the molecular weight determination of polar molecules of masses as high, in favourable cases, as 100 000 Daltons, and even, in exceptional hands, up to 1 000 000 Daltons). In FIB and LD, the sample is given a large pulse of energy, of which a relatively large proportion goes into translational, as opposed to internal, modes. The weak intermolecular bonds such as hydrogen bonds, which bond the analyte to its neighbours either in the solid state or in solution, are broken in preference to strong covalent bonds, and the sample is thrown out of its environment into the gas phase. In ESI, the large molecules to be analysed are 'carried along' into the gas phase by the evaporation of a solution of the large molecules into the gas phase in the presence of a strong electric field.

4.3.1 Fast ion bombardment (FIB or LSIMS)

In FIB desorption, the energy is provided by a beam of ions (most commonly Cs^+) of high translational energy, typically several keV. In an older technique, which was called fast atom bombardment (FAB), a beam of atoms was used rather than ions, but atoms are rarely used today. (FIB spectra are still sometimes called FAB.) Typically, a solution of a few micrograms of the sample is dissolved in a few microlitres of glycerol [$HOCH_2CH(OH)CH_2OH$] acting as a matrix of low volatility. A schematic illustration of the FIB source is given in Fig. 4.3.

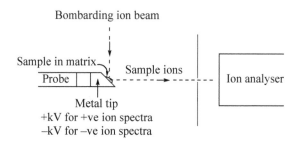

Fig. 4.3

When the Cs^+ ions strike the solution of the sample, the sample is desorbed, often as an ion, by momentum transfer. It is for this reason that the name LSIMS is also used for this technique, since the **S**econdary **I**ons are desorbed from a **L**iquid matrix. Neutral molecules M will also be desorbed, but since the polar molecules being analysed by FIB usually have relatively acidic (e.g. —CO_2H) or basic (e.g. —NH_2) functional groups, the corresponding ions (—CO_2^- or —NH_3^+) are also desorbed. The prevailing electric fields in the source then ensure that only the ions are efficiently transmitted into the analyser.

The beam only penetrates about 10 nm into the matrix. It is therefore helpful if the sample is marginally less hydrophilic than the matrix, so that the sample is concentrated near the surface. On the other hand, it should be in solution to avoid clumping. For both reasons, therefore, there is a need to have a range of matrices in order to find one that suits the sample. These include a thioglycerol-diglycerol mixture (1:1), tetragol [$HO(CH_2CH_2O)_4H$] and teracol [$HO(CH_2CH_2CH_2CH_2O)_nH$], which are successively less hydrophilic than glycerol itself.

4.3.2 Laser desorption (LD) and matrix-assisted laser desorption (MALDI)

In this ionisation method, a large energy pulse is passed to the sample from a laser. The sample molecule is thereby induced to leave its solid or liquid environment within a time of the order of 10^{-12} s. This time is too short to achieve an equilibrium distribution of the energy. Thermal decomposition is thereby reduced, or avoided altogether, in spite of the large amount of energy used. Efficient and controllable energy transfer to the sample requires resonant absorption of the molecule at the laser wavelength, either in the UV, which can induce electronic excitation, or in the IR, which can excite vibrational states. Typically, the laser pulses are applied for times in the range 1-100 ns.

The version of the technique that is most often used, especially for large peptides, proteins, oligonucleotides and oligosaccharides, is matrix-assisted laser desorption (MALDI). In this method, a low concentration of the sample is embedded either in a liquid or a solid matrix (molar ratio ranging from 1:100 to 1:50 000), which is selected to absorb strongly the laser light. A list of suitable matrix compounds is given in Table 4.1. In addition to these matrices, α-cyano-4-hydroxycinnamic acid (a UV absorbing matrix) is most commonly used in the analysis of peptides—a burgeoning field because of the importance of MS in the study of proteins, from which the peptides are often derived.

Table 4.1 Matrices for MALDI mass spectra

Matrix	Form	Usable wavelengths
Nicotinic acid	Solid	266 nm, 2.94 μm, 10.6 μm
2,5-Dihydroxybenzoic acid	Solid	266 nm, 337 nm, 355 nm, 2.79 μm, 2.94 μm, 10.6 μm
Sinapinic acid	Solid	266 nm, 337 nm, 355 nm, 2.79 μm, 2.94 μm, 10.6 μm
Succinic acid	Solid	2.94 μm, 10.6 μm
Glycerol	Liquid	2.79 μm, 2.94 μm, 10.6 μm
Urea	Solid	2.79 μm, 2.94 μm, 10.6 μm
Tris buffer (pH 7.3)	Solid	2.79 μm, 2.94 μm, 10.6 μm

The energy absorbed by the matrix is transferred indirectly to the sample, which reduces any sample decomposition. The matrix is chosen to have solubility properties similar to those of the sample, in order that the sample molecules are properly dispersed. Higher molecular weight oligomeric 'clumps' are produced as $2M^+$, $3M^+$, and so on, but these are usually minor components of the spectrum if a well-matched matrix is chosen.

4.3.3 Electrospray ionisation (ESI)

An 'electrospray' is the term applied to the small flow of liquid (generally 1-10 μl/min) from a capillary needle when a potential difference typically of 3-6 kV is applied between the end of the capillary and a cylindrical electrode located 0.3-2 cm away (Fig. 4.4). The liquid leaves the capillary as a fine mist at or near atmospheric pressure, and consists of highly charged liquid droplets. The charge on these droplets may be selected as positive or negative according to the sign of the voltage applied to the capillary. ESI is especially useful since it can be used to analyse directly the effluent from an HPLC column.

The use of a 'sheath' gas or 'nebulising' gas promotes efficient spraying of the solution of the sample from the capillary. Sample molecules dissolved in the spray are released from the droplets by evaporation of the solvent. This evaporation is accomplished by passing a drying gas across the spray before it enters a capillary. As the droplets are multiply charged, and reduced in size by evaporation, the rate of desolvation is increased because of repulsive Coulombic forces. These forces eventually overcome the cohesive forces of the droplet, and an MH^+ (or $M - H^+)^-$ molecular ion free of solvent is finally produced. The charged particles are carried, by an appropriate electric field, through a capillary and into an ion analyser. A typical

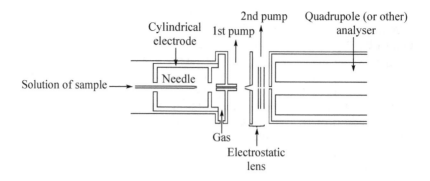

Fig. 4.4

droplet will contain a large number of sample molecules, and the separation of these, avoiding the observation of a polymeric aggregate, is also promoted by charge-charge repulsion. Solvents used in ESI are typically water/methanol mixtures; and frequently, in the case of positive ion work, traces of a volatile organic acid such as formic acid. ESI is a highly sensitive and powerful technique for the production of molecular ions from large polar molecules.

4.4 Ion analysis

Once ions have been obtained in the gas phase, they are repelled into an ion analyser, in order to be separated according to their masses, or, to be more precise, according to their m/ze ratios. This section summarises the main methods of ion analysis, some of which provide a medium level of mass resolution, say, up to a limit of 1 part in 10 000; others are good enough to give high resolution up to 1 part in 10^6.

4.4.1 Magnetic analysers

If we wish merely to separate ions of relatively small mass, for example to resolve m/z 210 from m/z 211, where these values represent singly charged fragments whose atomic constituents add up to masses of 210 and 211, respectively, it is sufficient to deflect the ions using only a strong magnetic field, as in Fig 4.5, but without the electrostatic sector that is illustrated there. In a magnetic sector, ions of larger mass are deflected less than ions of smaller mass, according to Eq. 4.1; where B is the strength of the magnetic field, r is the radius of the circular path in which the ion is travelling; and V is the potential used to accelerate the ions when they leave the source, determining the velocity of an ion when it enters the analyser.

$$\frac{m}{z} = \frac{B^2 r^2}{2V} \tag{4.1}$$

In Fig. 4.5, the poles of the magnet that provides the magnetic field of the magnetic sector lie above and below the plane of the paper. At specific values of B and V, paths of different radius r are followed for each value of m/z. By scanning the magnetic field B, Eq. 4.1 can be satisfied for ions of all m/z ratios, given the fixed values of r and V.

Magnetic analysers with a mass range of $\leq1,000$ Daltons may be used with EI (as shown in Fig. 4.5) and CI sources, since these ionisation methods are restricted to molecules with molecular weights in this mass range. Larger mass ranges (up to about 4,000 Daltons) are required when a FIB source is employed. Protonated oligomers of the FIB matrix $[(glycerol)_n H^+, m/z \ (92n + 1)]$ are often desorbed along with the sample ions, and are useful in calibrating the mass spectrum.

4.4.2 Combined magnetic and electrostatic analysers—high-resolution mass spectra (HRMS)

Higher mass resolution is achieved by passing the ion beam through an electrostatic analyser before it enters the magnetic sector (Fig. 4.5). In such a double-focusing mass spectrometer, ion masses can be measured with an accuracy of about 1 p.p.m. However, precise mass measurement requires the use of narrow source exit and collector slits, with a consequent loss in sensitivity.

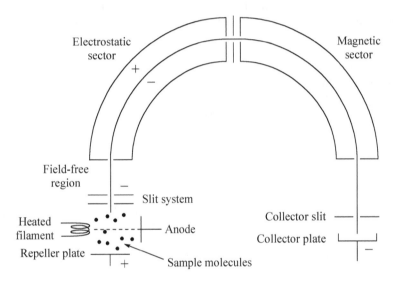

Fig. 4.5. The circular path followed by ions of a specific m/z value is indicated, for which the values of V and B are appropriate to let these ions pass through the collector slit.

When the molecular ion ($M^{\bullet+}$ or MH^+) is detectable, high resolution makes it possible to determine the molecular formula. Ions with the same nominal integral mass possess different exact masses, because the individual isotopes do not have exact integral masses. Thus, although CO, N_2, and $CH_2{=}CH_2$ all have the same integral mass (28), the exact masses of the three molecules are different: based on the convention that the atomic weight of ^{12}C is exactly 12, CO is 27.9949, N_2 is 28.0061, and $CH_2{=}CH_2$ is 28.0313. A mass spectrometer set for high resolution is accurate to about 1 p.p.m. for ions with m/z values up to 100 Daltons, and is therefore easily able to distinguish between these values. Furthermore, the level of accuracy, although it falls off with increase in mass, still makes it possible to determine unambiguously the molecular formula of many organic compounds with molecular weights up to about 1000 Daltons.

4.4.3 Ion cyclotron resonance (ICR) analysers

In ICR, a pulse of ions is injected into a cell at low translational energy, where a uniform magnetic field B constrains them into a circular path perpendicular to the direction of the magnetic field. For a singly charged ion, the frequency ω_c (the number of turns round the circle each second) is given by Eq. 4.2.

$$\omega_c = \frac{eB}{m} \tag{4.2}$$

If an alternating electric field of frequency ω_i is applied normal to B, an ion will absorb energy when $\omega_c = \omega_i$. Thus, an ICR mass spectrum can be obtained by fixing B and scanning ω_i so that ions of different mass successively satisfy the equation. The absorption of energy by the ions at resonance is measured using an oscillator detector similar to that used in NMR instruments. Better still, a spread of frequencies may be generated by using a pulsed radio-frequency electric field, and the spectrum analysed by Fourier transformation (FT). For high sensitivity, the ions are kept resonating for times in the range of milliseconds-seconds, necessarily at low operating pressures, typically 10^{-6} Nm^{-2}.

FT-ICR has the advantage of high sensitivity and high resolving power—in a conventional 5 cm cubic cell, as few as 10 ions can be detected. Since ion detection is non-destructive, signal detection can in principle continue for long periods, improving the signal-to-noise ratio. Ultra-high mass resolution can be achieved for small ions (e.g. $m/\Delta m = 1,000,000$ for m/z 100), but since resolving power decreases linearly with increasing mass, a drop in resolution to $m/\Delta m = 10\ 000$ can be expected for an ion of m/z 10,000.

4.4.4 Time-of–flight (TOF) analysers

One of the most commonly used methods of analysis for MALDI spectra is called *time-of-flight* (TOF). MALDI-TOF instruments are linear, bench-top devices, smaller than the dual-processing magnetic and electric sector instruments in which the stream of ions takes a curved path. The resolution of TOF instruments is improved by the use of a device known as a reflectron. Additionally, TOF ion analysis is very sensitive because there is no necessity to scan the spectrum. The problem with a scan is that the ions are not being recorded for a large fraction of the time in which they are being produced. In the general case, ions with charge z from MALDI are accelerated through a potential difference V, and thereby acquire a translational energy zeV. The relationship between the time of flight t and the distance d travelled through a field-free tube to reach the detector is given by Eq. 4.3.

$$t^2 = \frac{m}{z}\left(\frac{d^2}{2Ve}\right) \tag{4.3}$$

For simplicity, consider all those ions with a single charge, and therefore possessing a translational energy eV. In a given instrument, where the term in brackets in Eq. 4.3 is kept constant, the time taken for an ion of mass m to reach the detector is proportional to $\sqrt{m}$, with ions of largest mass having the lowest velocities and the longest time of flight over a given distance. Since mass analysis is achieved through the accurate measurement of time, the ions must be produced in pulses.

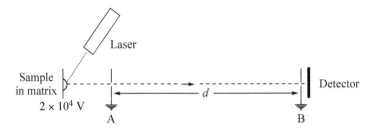

Fig. 4.6

A voltage drop from the source plate to the slit A (Fig. 4.6) accelerates the ions thrown out by the laser beam. The ions pass into the field-free flight tube between A and B. Since it is common to have differences in arrival times between successive mass peaks of $<10^{-7}$ s, fast electronics are required to distinguish successive peaks, but the time taken by the laser pulse can be made small compared to the differences in arrival times between successive spectra.

4.4.5 Quadrupole analysers

The arrangement of electrodes in a quadrupole mass filter is given in Fig. 4.7. A voltage U, typically 500-2000 V, and a radio-frequency potential V, oscillating from −3000 to +3000 V, are applied between opposite pairs of four parallel rods, which are 0.1-0.3 m long in most commercial instruments. Ions are injected along the z direction, along which they maintain a constant velocity. They travel with a wave pattern in the x and y directions (mutually orthogonal to each other, and also to z), controlled by the fluctuating potentials on the rods, so that under any given set of conditions ions of only one m/z value arrive at the detector, the others being captured by the rods. All ions can successively be brought to the detector by varying the amplitude of U and V, while keeping the ratio U/V constant, or by varying the radio-frequency potential V.

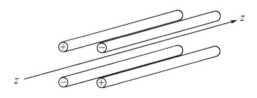

Fig. 4.7

Quadrupole mass spectrometers are powerful in the range up to m/z 1000, and increasingly even up to m/z 4000—especially where high resolution is not required and where simplicity of operation is an advantage. They are relatively small, and they are especially useful where GC and/or HPLC instruments are directly coupled to the mass spectrometer (although magnetic sector instruments can be made to perform these functions as well). Where the ion sources coupled to the instruments produce multiply charged ions (ESI and MALDI), their mass range extends up to z times 4,000 Daltons, making them very powerful in the analysis of molecules of biological origin.

4.4.6 Ion-trap analysers

Ion-trap mass spectrometers use electrodes to trap gas-phase ions in a small volume. They have the advantages of being relatively compact and, through being able to trap and retain ions, of high sensitivity. The two most common types of ion traps used in modern mass spectrometry are the quadrupole ion trap and the Orbitrap (introduced in 2005).

In the quadrupole ion trap, all ions created over a given time period are trapped inside the analyser through appropriate voltages applied to three electrodes which surround the trapping region. The ions are then sequentially ejected (according to their m/z values) towards a conventional electron multiplier detector, this sequential ejection being achieved by gradually changing the potentials applied to the electrodes.

In an Orbitrap mass spectrometer, ions are trapped because their electrostatic attraction to a central electrode is balanced by the centrifugal forces that act on the ions as they orbit round this central electrode. As in an FT-ICR mass spectrometer, the mass spectrum is obtained after Fourier transformation. The Orbitrap has a resolving power that decreases as the m/z value increases but, at the relatively high masses that are of interest in the field of proteomics, is still capable of a resolving power of <2 p.p.m. Thus, when used in conjunction with electrospray ionisation (ESI), the Orbitrap is important in the analysis of proteins (see later).

4.5 Structural information from mass spectra

4.5.1 Isotopic abundances

The masses and natural abundances of some isotopes which are important in organic mass spectrometry are given in Table 4.4 at the end of this chapter. All singly charged ions in the mass spectra of compounds which contain carbon also give rise to a peak at one mass unit higher, because the natural abundance of ^{13}C is 1.1%. For an ion containing n carbon atoms, the abundance of the isotope peak is $n \times 1.1\%$ of the ^{12}C-containing peak. Thus, the molecular ion for nonane ($C_9H_{20}^{\cdot +}$) gives an isotope peak at 129, one mass unit higher than the molecular ion, with an approximate abundance of 10% (9 × 1.1) of the abundance at m/z 128. Obviously, larger molecules with a lot of carbon atoms give much more prominent ($M^{\cdot +}+1$) ions. In the case of small molecules, the probability of finding two ^{13}C atoms in an ion is low, and ($M^{\cdot +}+2$) peaks are accordingly of insignificant abundance. Conversely, $M^{\cdot +}+2$, and even $M^{\cdot +}+3$, $M^{\cdot +}+4$, etc. peaks do become important in very large molecules (see later for details).

Although iodine and fluorine are monoisotopic, chlorine consists of ^{35}C and ^{37}Cl in the ratio of approximately 3:1, and bromine of ^{79}Br and ^{81}Br in the ratio of approximately 1:1. Molecular ions (or fragment ions) containing various numbers of chlorine or bromine atoms therefore give rise to the patterns shown in Fig. 4.8 (all peaks spaced 2 mass units apart). Obviously, the isotope patterns to be expected from any combination of elements can readily be calculated, and they provide a useful test of ion composition in those cases where polyisotopic elements are involved. Most of the remaining elements in Table 4.4 are essentially monoisotopic, with the exception of sulfur and silicon.

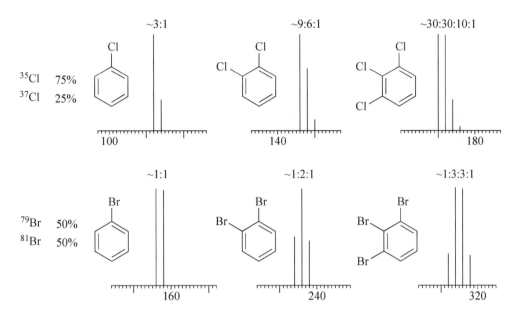

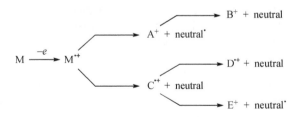

Fig. 4.8

4.5.2 EI spectra

Electron impact imparts so much energy to a molecular ion that it usually fragments. Because mass spectroscopy can be applied to extraordinarily small samples, it is the method of choice for the analysis of such trace materials as pesticide residues and drugs and their metabolites, especially in forensic and medical research; it is important in bringing cheating athletes to heel! The fragmentation pattern can also be used in structure determination, since the bonds that break are in general recognisably weak bonds, leading to the lowest energy combination of fragment ions and expelled neutrals.

The molecular ion produced by EI has an unpaired electron (it is a radical-cation, $M^{\bullet+}$). The molecular ions fragment in the ionisation chamber either by loss of a radical to give A^+, and/or by loss of a molecule with all its electrons paired to give another radcial cation $C^{\bullet+}$). Subsequently, these charged fragments (cations or radical-cations, respectively) may break down again in an analogous manner:

Note that once an ion lacking an unpaired electron has been generated, any further fragmentation occurs *via* loss of a neutral molecule, and not *via* loss of a radical. The bonds breaking, and the extent of fragmentation in the ionisation chamber, are dependent upon the chemical structure; stable $M^{\bullet+}$ ions lead to abundant molecular ions in the mass spectrum. Conversely, unstable $M^{\bullet+}$ ions can result in an absence of a molecular ion.

Fragmentation gives rise to a *pattern* of fragment ions, through which the compound may be characterised, especially if it is a previously known compound and the mass spectrum is already available for comparison. A typical result produced by the spectrometer is shown in Fig. 4.9, which is the EI spectrum of n-nonane **1**. The most prominent peak is called the *base peak*, and this defines the intensity 100%. In this case it comes at *m/z* 43. The *molecular ion* $M^{\cdot+}$ is the peak at *m/z* 128, and it has an intensity of 8% of the intensity of the base peak. Other prominent fragment ions can be seen at *m/z* 99 (5%), 85 (28%), 71 (22%), 57 (68%), 41 (42%), 29 (37%) and 27 (31%).

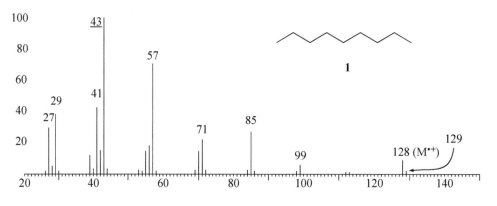

Fig. 4.9

If we did not know what this molecule was, the $M^{\cdot+}$ value of 128 could have been produced by any of the molecules $C_{10}H_8$, C_9H_{20}, C_9H_6N, C_9H_4O, $C_8H_{18}N$, and several other combinations with fewer and fewer carbon atoms. These formulae would have accurate molecular weights of 128.0625, 128.1564, 128.0500, 128.0262 and 128.1439, respectively. High resolution MS distinguishes between these (and other) possibilities. Typically the measured value might be $M^{\cdot+}$ = 128.1568, and would therefore identify the sample unambiguously as one of the nonanes C_9H_{20}. This information is arguably the single most valuable use of the mass spectrum in everyday structure determination. Spectrometers set to determine high-resolution spectra are equipped with a computer that automatically generates acceptable formulae for the ions it detects, and the chemist can read the formulae directly on the computer's print-out.

A high-resolution mass spectrum (HRMS) has largely replaced combustion analysis for determining molecular formulae, but it is wise to exercise caution before abandoning combustion analysis altogether. The exact mass measurement works whether the compound is pure or not, and it is easy to accept the existence of a peak with the expected exact mass as proving the formula, when all it does is prove that there was some of that compound present. Combustion analysis has different problems—it can give wrong answers when the compound is not pure, or when there is error in the analysis. It only suggests a possible empirical formula, and it is not capable of assessing the presence or measuring the amount of many elements now commonly found in organic structures.

Although there are some reliable fragmentation pathways induced in EIMS, not all yield to a simple analysis, and structure determination using mass spectroscopy is not an exact science. In the following sections, the spectra of some organic molecules are considered according to the functional groups they contain (or, in the case of hydrocarbons, do not contain). It is important to remember that you should not expect always to be able to account for every peak in a mass spectrum.

Aliphatic hydrocarbons. Let us begin with the spectrum of n-nonane **1** already seen in Fig. 4.9. It is convenient to picture the weakening, by loss of an electron, of many of the C-C bonds in $M^{\bullet+}$. Then, given sufficient vibrational energy, a bond like the one illustrated in the representation **2** breaks to give an ion **3** and a radical **4**.

2	**3**	**4**
	a cation C_5H_{11} *m/z* 71	a radical, no charge, not detected

The ion **3** with a molecular formula of $C_5H_{11}^+$ gives rise to the peak at *m/z* 71. The radical **4** with a molecular formula of $C_4H_9\bullet$ has no charge, it is not deflected, and does not appear in the spectrum. In this way, numerous C_nH_{2n+1} cations can be generated, giving the ion series *m/z* 99, 85, 71, 57 and 43. The lower mass ions of this series may not only be formed directly, but also by the loss of ethylene **6** from one with higher mass:

3	**5**	**6**
C_5H_{11} *m/z* 71	a cation C_3H_7 *m/z* 43	neutral, not detected

A useful way of conveying the structural features responsible for such ions is shown in the drawing **7**, where the wavy lines identify the bonds broken and the numbers, and the side on which they are placed, identify the mass-to-charge ratios of the cations produced. Note that the fragmentation of any of the C—H bonds is unfavourable, because hydrogen atoms and unsolvated protons are very high in energy.

7

Ions of the general formula $C_nH_{2n-1}^+$ form a second series of fragment ions. For example the ions at *m/z* 27, 41 and 55, which occur two mass units below the more prominent ions for the ethyl, propyl and butyl cations at 29, 43 and 57. Their formation occurs by loss of a saturated hydrocarbon molecule or, less commonly, H_2 from ions of the $C_nH_{2n+1}^+$ series, and illustrates the generalisation that once the unpaired electron has left the molecule, the fragmentation that takes place is normally by loss of a neutral molecule, rather than of a radical:

$$C_4H_9^+ \rightarrow C_3H_5^+ + CH_4$$

Let us look now at an isomer of n-nonane, namely 3,3-dimethylheptane **8**. The spectrum is given in Fig. 4.10, where we see that most of the fragment ions are similar to those for n-nonane in Fig. 4.9. But there are subtle differences, and a not so subtle difference—the absence of the molecular ion at m/z 128.

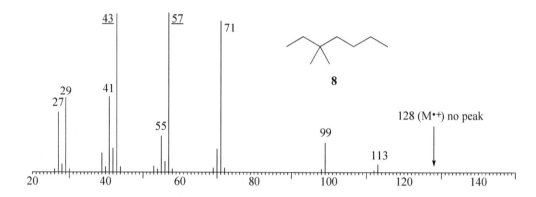

Fig. 4.10

The molecular ion is absent because there are now several energetically more favourable fragmentation pathways, and therefore the parent ion does not survive long enough to escape the ionisation chamber. One of the more subtle differences is the increase in the relative intensities of the peaks at m/z 99, and 71, produced by the fragmentations **8** and **9**. These ions carry a higher proportion of the ion current because they are tertiary, and therefore even lower in energy (better stabilised by alkyl substitution) than the primary cations produced in the spectrum of n-nonane. A second subtle difference is that the loss of the unstabilised methyl radical in the fragmentation **10** is more prominent, because the formation of the tertiary cation **11** compensates for the high energy of the accompanying methyl radical. The third subtle difference is that the ion at m/z 57 is also increased in intensity. In this case the fragmentation **9**, with the charge kept by the relatively unstabilised n-butyl cation, is compensated for by the formation of the well-stabilised tertiary pentyl radical.

From the differences between Figs. 4.9 and 4.10, we learn that *the preferred fragmentations are those where the sums of the energies of the fission products are at a minimum* (as is indeed demanded by theory). Tables 4.5 and 4.6 at the end of this chapter give the enthalpies of formation of common radicals and ions, which are useful to identify preferred sites of fragmentation.

Identifying the molecular ion. In EI spectra, the absence of a molecular ion (as in Fig. 4.10) is not uncommon, and it is often a problem to decide whether the peak at highest mass is a molecular ion or not. In Fig. 4.10, the peak at *m/z* 113 could not be the molecular ion *because it is at an odd number*. In this molecule, the only elements are C and H, and all neutral hydrocarbons have an even molecular weight. If the molecule has one or more oxygen or silicon or sulfur atoms, the molecular ion will still be an even number. The same is true with one or more halogens replacing the same number of hydrogen atoms, since all the abundant isotopes of the halogens have odd atomic weights. But if there is a single nitrogen atom, the molecular ion will register as an odd number. Triethylamine is $C_6H_{15}N$, with a molecular weight of 101. More generally, if the neutral molecule has an odd number of nitrogen atoms, the molecular ion will be an odd number, and if it has an even number of nitrogen atoms, the molecular ion will be an even number. In summary, a molecule composed only of one or more of the elements C, H, O, Si, S and the halogens will have an even molecular weight. A molecule having an odd number of nitrogen atoms will have an odd molecular weight.

Another useful criterion for identifying a molecular ion is that, if there are any peaks corresponding to the loss of 3-14 mass units from the highest recorded ion, that ion is probably not the molecular ion. Fragmentations corresponding to the loss of H_3 to H_{11} and of C, CH or CH_2 do not typically occur. Such ions, if they are present, must normally be produced by fragmentations of an ion that is greater in mass than the highest recorded ion. Thus, the gap of 14 mass units between the ion at *m/z* 99 and the highest recorded ion at *m/z* 113 in Fig. 4.10 is not likely to represent the loss of a CH_2 group. Its presence is a further indication that the ion at *m/z* 113 is not a molecular ion.

The control of fragmentation by functional groups. We now look at the ways in which aromatic rings, and some functional groups, influence fragmentation. In these molecules, the highest energy molecular orbital is frequently a non-bonding atomic orbital on the hetero-atom, or a π-bonded molecular orbital, and it is useful to consider $M^{\cdot+}$ to correspond to the species generated by loss of one of these pairs of electrons.

Cleavage adjacent to an aromatic ring. This is frequently a preferred cleavage. When the ring is benzene, it is called a *benzylic cleavage*. A benzene ring can lower the energy of a radical or a cation adjacent to it, by delocalisation of the odd electron or the empty orbital. The cation **13** is called the benzyl cation. The stabilising effect of delocalisation makes the cleavage **12** of benzylic bonds easier than it would otherwise be, and this is helped by the relative ease with which an electron is removed from the π-bonds of the benzene ring in the first place. Cleavage of a benzylic substituent R is a common pathway. Even a hydrogen atom, in spite of its high energy, can be cleaved off toluene **14** because there is no easier pathway, as seen in the mass spectrum of toluene in Fig. 4.11, where the benzyl cation **13** gives the base peak at *m/z* 91.

12   **13**

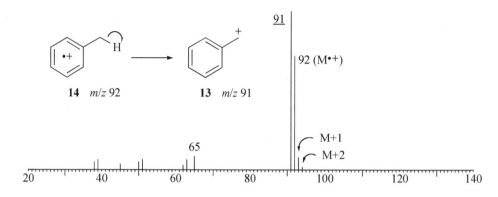

Fig. 4.11

Because the loss of a hydrogen atom is relatively unfavourable, the molecular ion is more abundant than usual, making the M+1 and M+2 ions easy to see. The benzyl cation **13** is probably the appropriate structure for *m/z* 91 when it is initially generated, but there is evidence that it can rearrange to the aromatic tropylium cation **15** before it fragments further. Such rearrangement is possible because the internal energy of **13** (or indeed of **15**) must be very high to enable it to fragment. Some further fragmentation of **15** occurs by the loss of a molecule of acetylene, probably to give the cyclopentadienyl cation **16**. In accord with the high-energy requirements for any fragmentation of **15**, there is no simple mechanistic pathway, yet, whatever the method of generating the benzyl cation, this sequence of events is always followed.

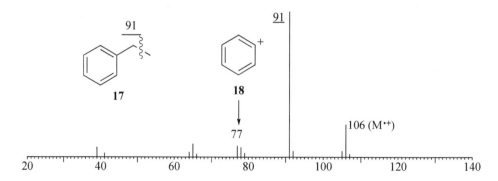

The cleavage of a benzylic alkyl group is much easier, as seen in the mass spectrum of ethylbenzene **17** (Fig. 4.12), where the molecular ion is less abundant than it is for toluene in Fig. 4.11. The phenyl cation **18**, on the other hand, is not stabilised by delocalisation of the positive charge, and the fragment ion *m/z* 77 is of low abundance.

Fig. 4.12

Cationic rearrangement is common in chemistry, and therefore also in the mass spectrometer, where it can take place counter-thermodynamically if it leads to structures from which a stable molecule can be easily lost. Thus, t-butylbenzene loses a methyl radical from $M^{\cdot+}$, and then loses a molecule of ethylene. Probably the initially generated cumyl cation $PhCMe_2^+$ rearranges to a 1-phenyl-n-propyl cation $PhCH_2CH_2CH_2^+$, from which ethylene can be lost to give a benzyl cation.

In disubstituted benzenes, there is always formally a choice between the fragmentation of one or the other side chains. Since the energetically easier cleavage is always preferred, it is possible to deduce which fragmentation will occur by measuring the energy requirements for the fragmentation of each side chain when present alone (Table 4.8). Thus, p-cyano-t-butylbenzene loses a methyl radical, rather than suffering the loss of HCN. On the other hand, a bromine radical is lost exclusively from the molecular ion of p-bromoaniline, because the cleavages involving the NH_2 group require higher energies. o-Disubstituted benzenes are sometimes anomalous in that fragments made up from parts of *both* substituents can be lost. Thus, o-nitrotoluene fragments to give a benzisoxazoline cation by loss of an OH radical, with the hydrogen atom coming from the methyl group and the oxygen atom from the nitro group.

All these features of the mass spectra of benzene compounds are also seen in other aromatic compounds and in heterocyclic aromatic compounds.

Heteroatom-based cleavage. Oxygen, sulfur, nitrogen and halogen atoms, often called heteroatoms, have one or more lone pairs of electrons. If they are present in the molecule, the lone pairs are less involved in bonding than are the other electrons, and are usually in the highest-energy occupied molecular orbital. One of them is therefore more easily expelled than an electron in a σ or π bond. The radical cations produced are initially localised on the heteroatom, and fragmentation is therefore easy by cleavage of the β-bond. We can see how this works in the spectrum of 2-butanol **19** in Fig. 4.13.

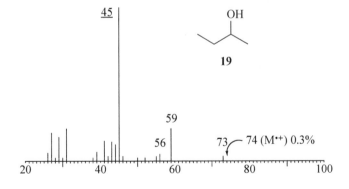

Fig. 4.13

The first-formed ion will be the radical-cation, with the odd electron localised in an orbital on the oxygen atom. This electron will pair with an electron from one of the three bonds β to the oxygen atom. These are the C—H bond, the bond between C-2 and C-1, and the bond between C-2 and C-3. The other electron in these bonds will be localised on a hydrogen atom, or a methyl radical, or an ethyl radical. Thus, the fragments carrying the current will be the protonated butanone **20**, the protonated propanal **21**, and the protonated acetaldehyde **22**, respectively, and the relative intensities of the peaks at *m/z* 73, 59 and 45 largely reflect the relative stabilities of the radicals that are lost.

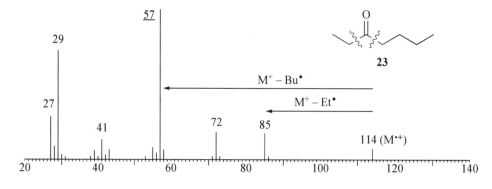

A rather characteristic peak from alcohols is that from the loss of water (M − 18), which gives a peak at *m/z* 56 in this case. Note that this is unusual in being a fragment with an even molecular weight—it is one of the comparatively rare cases where a stable *molecule* is lost from the radical-cation, and it occurrs because of the thermodynamic stability of water. Tertiary alcohols almost never show a molecular ion, and usually show an abundant M − 18 ion, but primary alcohols are more apt to lose H_2 to give the aldehyde as the first fragmentation.

The fragmentation of simple ketones follows the same pattern—the electron ejected is one of those from one of the lone pairs on the oxygen, and the radicals expelled are those that are β to the oxygen. Thus, in the mass spectrum of 3-heptanone **23** in Fig. 4.14, the radicals that are cleaved off are the butyl and ethyl radicals.

Fig. 4.14

The nomenclature, however, is confusing because the bond that is cleaved is that between the carbonyl group and the carbon that is called the α carbon, when we are naming ketones. As a result this cleavage is called α cleavage, even though the bonds cleaved in 3-heptanone **23** bear the same relationship to the oxygen atom as the bonds cleaved in an alcohol like 2-butanol **19**.

The peak at m/z 29 is the ethyl cation produced by loss of CO from the ion **25**. A similar loss of CO from the ion **24** gives a boost to the abundance of the ion at m/z 57, because the butyl cation and the propionyl cation **25** have the same integral mass. High resolution would reveal that the ion at m/z 57 is actually composed of two closely spaced peaks. The ion at m/z 41 is an allyl cation, produced by a fragmentation typical of hydrocarbons, as seen in the spectrum of nonane.

There is another abundant ion, at m/z 72, which is again a relatively unusual case of a fragment having an *even* molecular weight produced by the loss of a *molecule* from $M^{•+}$. It corresponds to the radical-cation of the enol **27**, and arises by a McLafferty rearrangement **26**, in which propene **28** is lost by transfer of the γ-hydrogen atom from carbon to oxygen. For obvious reasons it is also called β cleavage.

This fragmentation is useful, because it is diagnostic of a carbonyl group having a γ-hydrogen atom. The m/z value of the peak and the mass of the fragment lost can help to identify the substituents, if any, attached to the two halves of a ketone. Both the α-cleavage and the McLafferty rearrangement are reminiscent of the Norrish Type I and Type II fragmentations seen in the photolysis of ketones. Similar fragmentations are also seen with other carbonyl compounds, and Table 4.11 at the end of this chapter lists the m/z values for the ions produced from the most simple member of each of the common carbonyl compounds.

Amines even more easily undergo the loss of one of the lone pair of electrons, followed by cleavage of the bonds β to the heteroatom. The amine **29** has three such bonds to alkyl groups with this relationship, and they give rise to the most prominent peaks in the mass spectrum shown in Fig. 4.15. This time, the molecular ion at m/z 129 is an odd number, and consequently all the major fragments are even-numbered, characteristic patterns for a molecule containing an odd number of nitrogen atoms.

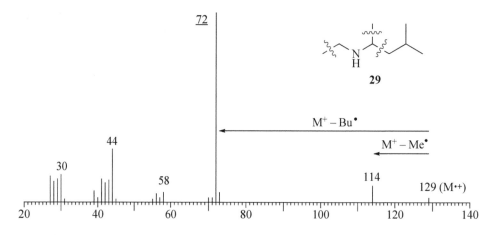

Fig. 4.15

There are two ways that a methyl radical can be lost, **30** and **32**, and one for an isobutyl radical to be lost **34**. The latter is the major pathway, because the larger radical is more stable than the methyl radical. Each of the cations **31** and **35**, can lose a molecule of an alkene, which accounts for the formation of the fragments at *m/z* 30 and 44. The loss of any of the three β-hydrogen atoms is, as usual, less favourable than the loss of the alkyl radicals, and there is no significantly abundant M − 1 peak.

The halogens, although they do have lone-pairs, give rise to a different pattern of fragmentation—primarily because halonium ions are much less stable than are imminium or oxonium ions. However, the halogens do form stable radicals. Provided that the loss of a halogen atom does not give rise to a vinyl or phenyl cation, which would have an unfavourable positive charge at a trigonal centre, it is likely to be observed. Thus, in the spectrum of n-butyl bromide **36** in Fig. 4.16, the loss of a bromine atom is the major fragmentation pathway. The minor fragmentations that retain the bromine atom in the

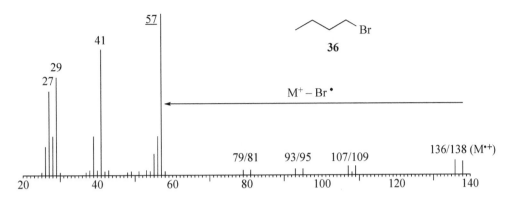

Fig. 4.16

charged fragment are recognisable because they give rise to two peaks of equal intensity two mass units apart. One of these is the bromonium ion itself **38** at 79/81, and another at m/z 93/95 is the result of β-cleavage **39**. Cleavage of the γ bond gives rise to the ion **40** at m/z 107/109. The ion at m/z 41 ($CH_2=CHCH_2^+$) arises by the loss of methane from m/z 57, the ethyl cation (m/z 29) from the loss of ethylene from the butyl cation **37**, and 27 ($CH_2=CH^+$) by loss of H_2 from the ethyl cation.

Note the useful generalisations that are affirmed in this section: the major fragmentation pathways from molecules with an even molecular weight take place by the loss of radicals to give charged fragments with odd molecular weights. Conversely, molecules with an odd molecular weight mainly give fragments with an even molecular weight. Recall that in EI mass spectra it is not normally profitable to attempt to identify the origins of all the peaks. At the end of this chapter, Tables 4.9, 4.14 and 4.15 may help in identifying some of the more common fragments, both those characteristic of common functional groups (Table 4.9), those broken off from the ions (Table 4.14), and those carrying the charge (Table 4.15).

4.5.3 CI spectra

The EI spectrum of the common plasticizer dioctyl phthalate **41** in Fig. 4.17a contains no peak corresponding to the molecular ion $M^{•+}$, but the CI spectra in Figs 4.17b and 4.17c give abundant MH^+ peaks at m/z 391. The spectrum using methane shows some fragmentation, but the use of isobutane as a CI gas, transfers very little energy into the analyte, and leads to very little fragmentation.

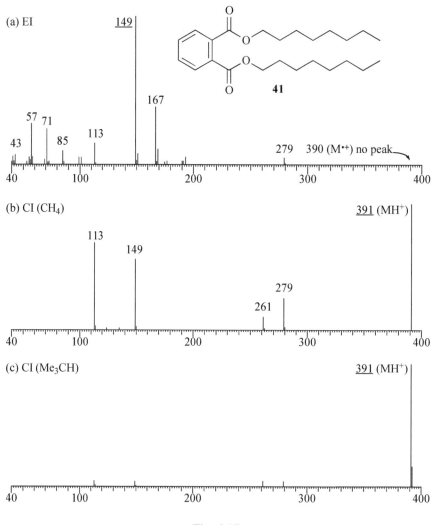

Fig. 4.17

The fragmentation in the CI spectrum in Fig. 4.17b can be rationalised by a McLafferty-like rearrangement of the protonated molecular ion **42** to give the ion **43**, and the formation of phthalic anhydride derivatives **44** and **45**, and the ion at m/z 113 is the octyl cation $C_8H_{17}^+$. The key lesson is that CI is a powerful method for the determination of the molecular weights of volatile compounds because, whatever the CI gas, the degree of fragmentation is much reduced relative to that seen in EI spectra.

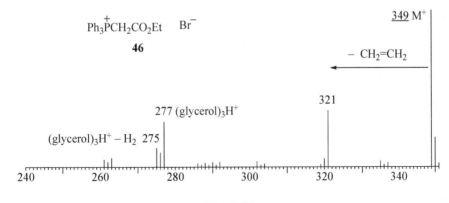

42 *m/z* 391

−H₂O

43 *m/z* 279

44 *m/z* 261

45 *m/z* 149

4.5.4 FIB (LSMIS) spectra

Both positive- and negative-ion FIB mass spectra can be obtained, although the latter method is often reserved for the analysis of compounds that form unusually stable anions (e.g. molecules containing sulfate groups). Molecular weights are normally available through the observation of abundant MH^+ ions in positive-ion spectra, and abundant $(M - H^+)^-$ ions in negative-ion spectra. FIB spectra frequently contain structurally useful fragment ions, but in the spectra of large molecules obtained from a matrix, MH^+ is usually the most abundant ion, making molecular weight determination easy by this method. It is not uncommon to observe $(M + Na)^+$ or $(M + K)^+$ ions in FIB spectra, from the presence of traces of the salts of these cations. Hence, it is also possible deliberately to create such ions by adding sodium or potassium salts to the matrix, where the presence of MH^+ and MNa^+ ions 22 mass units apart can help to identify the molecular ion.

$Ph_3\overset{+}{P}CH_2CO_2Et$ Br^-

46

$\underline{349}\ M^+$

$-\ CH_2{=}CH_2$

321

277 (glycerol)₃H⁺

(glycerol)₃H⁺ − H₂ 275

240   260   280   300   320   340

Fig. 4.18

FIB is used in the analysis of relatively involatile molecules up to molecular weights of *ca.* 4000 Daltons. Fig. 4.18 shows the FIB spectrum of the phosphonium bromide salt **46**, where the phosphonium cation *(m/z* 349) evaporated directly, and did not need to be protonated. The molecular ion is the most abundant ion, and, since it is not a radical, the only significant fragmentation is the loss of a molecule. Hydrocarbons, and other highly hydrophobic molecules such as steroids with little functionality (low basicity), are not handled well by particle bombardment methods.

Multiply charged ions such as MH_2^{2+} are occasionally seen in FIB mass spectra, especially when there are two basic residues (as for example in arginine) in a peptide, but such ions are normally of low abundance. The ions produced by FIB show relatively little fragmentation, but peptides having molecular weights in the range 300-3000 Daltons fragment to give valuable sequence information. Cleavages occur on either side of the nitrogen atom of peptide bonds, and rarely anywhere else to a significant extent. The fragment **47** is already charged, and appears only in positive-ion spectra. The fragments **48**, **49** and **50** are recorded in positive-ion spectra with an extra proton (H^+), and in negative-ion spectra with a missing proton.

The fragments **47-50** identify, residue by residue, the amino acids making up the peptide. Leu and Ile are isomers and have to be differentiated by amino acid analysis, but Lys and Gln, which have the same integral mass, can be differentiated by acetylation of the former. An example of peptide sequencing is illustrated by the negative ion spectrum of a peptide toxin in Fig. 4.19, which gives the partial sequence X-Ile-Asp-Asp-Glu-Gln. In conjunction with the positive-ion spectrum and the molecular weight, the total sequence was shown to be PhCO-Ala-Phe-Val-Ile-Asp-Asp-Glu-Gln. Since FIB mass spectra contain a small peak at essentially every mass, reliable mass calibration of the spectrum is not usually a problem.

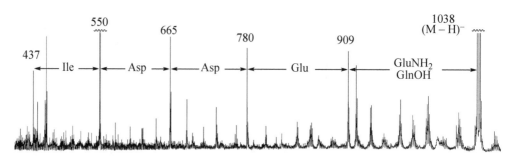

Fig. 4.19

In the interpretation of Fig. 4.19, and in the interpretation of such spectra in general, one looks for a series of fragment ions that are successively separated by amino acid mass differences. The atomic masses of the proteinogenic amino acids, which are used in this way in peptide sequencing, are given in Table 4.13 at the end of this chapter. A sequence can be confirmed by the self-consistent connection of a series of peaks, often supported by amino acid analyses of the pure peptides identifying which amino acids are present. Additional peaks which cannot be so interpreted can then be disregarded.

Another aid in determining the structure of polar molecules with many functional groups is to carry out on a microgram scale a reliable reaction, characteristic of a given functional group, preferably without tube-to-tube transfer, and measure the difference in molecular weight before and after reaction. The increase in molecular weight, often measured using FIB, indicates the number of such functionalities. Table 4.2 lists some useful reactions reliably giving high yields.

Table 4.2 Selective microscale reactions of some common functional groups

Functional group	Reagent	Product	Change in mass per functional group
RNH_2	Ac_2O/H_2O (30 min)[a]	RNHAc	+42
ROH	Ac_2O/pyridine (overnight)	ROAc	+42
RCO_2H	0.5% HCl/MeOH (overnight)	RCOOMe	+14
$RCONH_2$	$PhI(OCOCF_3)_2$	RNH_2	−28

[a]Reaction mixture buffered to pH ~ 8.5 with NH_4HCO_3.

4.5.5 MALDI spectra

The generation of ions by matrix-assisted laser desorption, and analysis of the resulting ions by a time-of-flight instrument (the MALDI-TOF combination) can give approximate molecular weights for biomolecules, even in the range 100 000-200 000 Daltons. However, the precision of the mass determination at these large masses may not be better than 10-100 Daltons. As a consequence, the addition of a proton to the molecule in the production of the gas-phase ion is not something that can be tested. Hence, the species analysed are usually referred to as 'M^+', even though it is likely that it actually corresponds to MH^+. The spectra obtained from many laser shots can be accumulated in order to give an improved signal-to-noise ratio. Using this technique, extraordinary sensitivity can be achieved, allowing 10^{-15} moles or less to be detected.

The power of the method is illustrated by the determination of the molecular weight of a monoclonal antibody (Fig. 4.20). Antibodies are moderately large proteins, produced by the immune system to bind to foreign substances that have appeared in the body. Normally, a given foreign substance (the antigen) causes the production of a mixture of such antibodies. However, if one particular cell of the activated set of cells in the immune system is cloned, then single (pure) antibodies can be produced. These *monoclonal antibodies* are of great importance in diagnostic medicine, since their binding to an antigen can be coupled to the generation of a colour response. Through such colour responses, the diagnosis of disease is greatly facilitated. The characterisation of monoclonal antibodies is therefore important. In the case illustrated in Fig. 4.20, all the prominent high molecular weight ions are either the molecular ion M^+ or are related to it, either as a simple multiple ($2M^+$), or as ions with more than one positive charge (M^{2+} and M^{3+}), or both ($2M^{3+}$ and $3M^{2+}$). The precision of the molecular weight determination is 149 190 ± 69 Daltons.

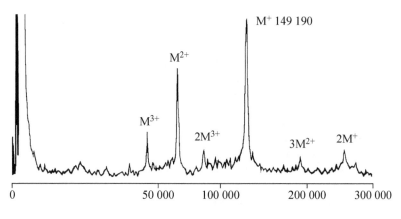

Fig. 4.20

4.5.6 ESI spectra

ESI typically produces multiply charged ions from biopolymers. Since the position of an ion in the mass spectrum is determined by the m/z ratio, multiple charges reduce the m/z value at which an ion of mass m appears in the spectrum. This property is of enormous importance in bringing the ions characteristic of large molecular weight species to a low m/z value (see also Fig. 4.20). It allows an ESI spectrum of a biopolymer to be measured by quadrupole analysers, even though these instruments were formerly considered to be suitable for the determination of molecular weights only up to about 4000. The principle is illustrated in the ESI spectrum in Fig. 4.21a of the peptide melittin **51** (a component of the sting venom of the honey bee, which is able to promote the bursting of cells by destabilising their membranes). Thus, ions characteristic of the molecular weight of 2846 Daltons occur even in the m/z region from 400 to 600. Similarly, the determination of the molecular weight (5064 Daltons) of human parathyroid hormone **52**, made up of a sequence of 44 amino acids, is readily possible using a quadrupole, as shown in Fig. 4.21b). In representing the structures **51** and **52,** the one-letter code for the amino acids is used (Table 4.13). The molecular weights (M_r) in the figures are average values including all isotopes.

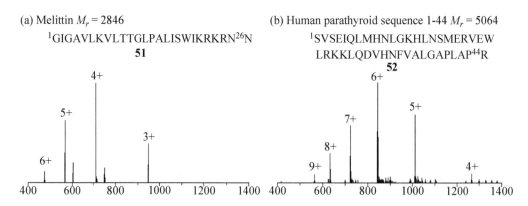

Fig. 4.21
(Redrawn with permission from *Anal. Chem.*, 1990, **62**, 882.)

The width of the distribution of charged states is often about half of the mass of the charged state with the highest mass, as it is in Fig. 4.21. The highest charged state often correlates quite well with the number of relatively basic functionalities in the amino acids. The side chains $—(CH_2)_4NH_2$ (pKa $ca.$ 8) in lysine, $—(CH_2)_3NHC(=NH)NH_2$ (pKa $ca.$ 12) in arginine, and imidazole (pK_a $ca.$ 6.5) in histidine, and the N-terminus of the peptide or protein, are relatively easily protonated to the corresponding positively charged ions. Thus, melittin has a total of 5 Lys, Arg and His residues, and human parathyroid hormone (1-44) has 9. Adding an extra charge from the protonation of the N-terminus leads to an expectation of maximum charged states of 6 and 10, respectively, close to the observed values of 6 and 9.

Since the number of basic residues in a protein is usually roughly proportional to the molecular weight, much larger molecules than these peptides can still be studied. Charged states of 40-80 would not be unusual in a protein with a molecular weight as high as 80 000 Daltons, and this brings the m/z values of the ions within the range covered by quadrupole analysers. Since the molecular weight information is contained in a number of the prominent peaks, as in Fig. 4.21, there is an algorithm to calculate the molecular weight using the data from all of them—permitting a determination with relatively high confidence and precision. Table 4.3 gives some examples of molecular weights determined in this way, along with the calculated molecular weights based on the weighted average of all the isotopes that are present (see Table 4.4 at the end of the chapter).

Table 4.3 Molecular weight determination of proteins by ESI-MS

Compound	M_r (measured)	M_r (calculated)	M_r error, %
Bovine insulin	5733.4	5733.6	−0.01
Ubiquitin	8562.6	8564.8	−0.02
Thioredoxin (*E. coli*)	11672.9	11673.4	−0.00
Bovine ribonuclease A	13681.3	13682.2	−0.01
Bovine α-lactalbumin	14173.3	14175.0	−0.01
Hen egg lysozyme	14304.6	14306.2	−0.01

The method is also one of remarkable sensitivity. Determinations of molecular weights of proteins can be made with the consumption of only $ca.$ 20 femtomoles (20×10^{-15} moles) of material. For a molecular weight of 10 000 Daltons, this corresponds to the consumption of only 200 picograms of material (less than one-thousandth of a microgram!).

Many mass spectrometry facilities operate ESI-TOF-MS combinations that may be used for automated high resolution measurements. These bench-top systems can provide <5 p.p.m. mass accuracy, and are therefore extremely useful in the determination of molecular formulae of molecules in the smaller molecular weight range.

4.5.7 ESI-FT-ICR and ESI-FT-Orbitrap spectra

When an ESI source is used in combination with FT-ICR, the resolution is much better than that of the ESI-quadrupole combination. However, FT-ICR instruments are more demanding to maintain, and require very low pressures to operate with their highest efficiency.

The spectrum of the protein cytochrome c is shown in Fig. 4.22, with the expected distribution of charged states. If the peak from any one of these states is expanded at high resolution, the peaks from the various combinations of ^{13}C isotopes are resolved, as shown in the detail in Fig. 4.22 of the cluster of peaks just above m/z 773. For a protein of this size (just over 12 000 Daltons) with several hundred carbon atoms, the isotope peaks from species containing from 0 to 8 ^{13}C atoms are quite abundant. Since the carbon isotope peaks for a singly charged ion must be separated by 1.0034 Daltons (Table 4.4), the number of these isotope peaks appearing within a single unit on the m/z scale defines the number of charges on the ion. The resolution of peaks separated by only 0.0625 units on the m/z scale at 773, from a molecule with an average molecular weight of 12 358.34 Daltons, emphasises the superb resolving power of ESI-FT-ICR.

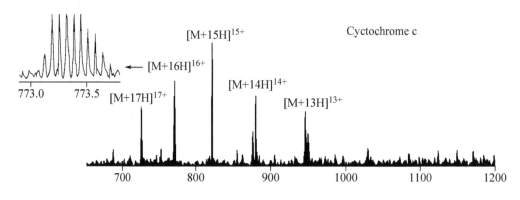

Fig. 4.22
(Redrawn with permission from *Proc. Natl. Acad. Sci. USA*, 1992, **89**, 286.)

4.6 Separation coupled to mass spectrometry

4.6.1 GC/MS and LC/MS

The separation and detection of components from a mixture of organic compounds is readily achievable by gas chromatography (GC), or by high-performance liquid chromatography (HPLC). Mass spectrometry, because of its high sensitivity and fast scan speeds, is a technique suited to the analysis of the small quantities of material eluted from these chromatographs. The association of the two techniques has, therefore, provided a powerful means of structure identification for the components of natural and synthetic organic mixtures, and for the rapid identification of the products of chemical reactions. Mass spectra of acceptable quality are potentially obtainable for every component that may be separated by the gas or liquid chromatograph, even though the components may be present in nanogram quantities and eluted over periods of only a few seconds.

When GC/MS and LC/MS were first introduced, the carrier gas (frequently helium) or the solvent had to be removed, especially when the ionisation chamber of the mass spectrometer was one of those operating at very low pressure. Although this is still true in EI-GC-MS work, modern technologies readily achieve the desired reduction in pressure. Additionally, CI sources are available which operate at atmospheric pressure (APCI sources), and such operation facilitates the coupling of both GC and LC instruments to the mass spectrometer, making it possible, for example, to analyse products directly from a reaction mixture. Perhaps most importantly, LC is readily coupled to the mass spectrometer where the LC effluent is directly used in an ESI source. The mass spectrometer is normally used as the detector, through measurement of the total ion current at any point in the chromatogram.

In LC-ESI-MS work, if the analyte contains groups that readily form ions (e.g. amino or carboxyl groups), and especially if the solvent includes a volatile buffer which can promote the formation of ions (such as formic acid or ammonium acetate), the droplets in the mist are to some extent positively or negatively charged. Solvent molecules are removed by fast pumping, leaving MH^+ or $(M - H^+)^-$ ions in the gas phase, where they can be extracted from the chamber by an electric field, and injected into the mass analyser. Good mass spectra of amino acids, peptides, nucleotides, and antibiotics up to molecular weights of around 2000 Daltons can be obtained fairly routinely.

Such is the power of GC/MS and LC/MS that the majority of mass spectrometers are today sold as integrated combinations of the separation and analysis techniques. Bench-top GC and LC/time-of-flight mass spectrometry systems have become available that offer sufficient mass accuracy to be used in the routine screening of the crude products of organic reactions and of complex biological samples. Although these instruments do not offer the highest mass resolution, they may be used to measure the 'monoisotopic mass' of ions to within 5 mDa (this is the mass of the isotopic peak which is composed only of the most abundant isotopes of those elements present). Thus, the monoisotopic mass can be calculated using the atomic masses of these most abundant isotopes. A further tool for identification is the degree to which the isotopic pattern of a detected compound agrees with the theoretical pattern of any analyte.

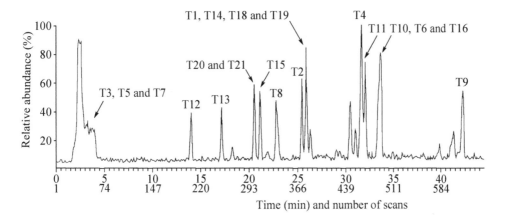

Fig. 4.23
(Redrawn with permission from *Anal. Chem.*, 1991, **63**, 1193.)

Because large amounts of data are collected from a GC/MS or LC/MS run, the system is usually coupled on-line to a computer-controlled data system (Sec. 4.8). The output from the computer shows the usual trace from a GC or HPLC, with the peaks identified by code letters and numbers, from which the mass spectra can be called up. For example, the methionyl human growth hormone is a protein, which was hydrolysed by digestion with the enzyme trypsin; this enzyme cleaves the protein at the C-terminal side of each of the basic residues lysine and arginine. The numerous peptides thus generated all contain two basic sites, namely the C-terminal Lys or Arg, and the N-terminus of the peptide. Figure 4.23 illustrates the HPLC trace obtained, and Figs. 4.24a and 4.24b show the ESI spectra obtained from two of the peaks, labelled T11 and T12. The mass spectra are relatively simple, containing molecular weight information only in MH^+ and MH_2^{2+} ions; they actually give the molecular weights of the peptides DLEEGIQTLMGR and LEDGSPR, which are contiguous in the growth hormone sequence.

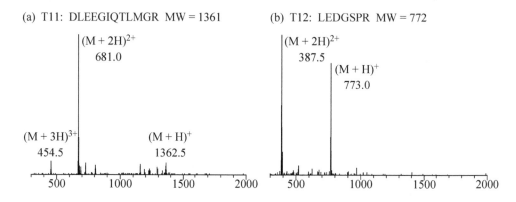

(a) T11: DLEEGIQTLMGR MW = 1361

(b) T12: LEDGSPR MW = 772

Fig. 4.24
(Redrawn with permission from *Anal. Chem.*, 1991, **63**, 1193.)

4.6.2 MS/MS

A mass spectrometer with both electrostatic and magnetic sectors (Fig. 4.5) can be used both for the separation of the various molecular ions produced from a mixture of components, and to obtain structural information from each of these molecular ions. However, in contrast to the arrangement shown in Fig. 4.5, this technology requires that the magnetic sector *precedes* the electrostatic sector.

To illustrate the principle, consider a mixture of three compounds (of molecular weights M_1, M_2 and M_3) that is ionised by one of the 'soft' ionisation techniques (i.e. CI, FIB, ESI, which afford abundant molecular ions). If the magnetic field is set to pass only M_2H^+, then only these ions will be directed into a collision chamber placed between the magnetic and electric sectors. An inert gas is introduced into this chamber at a moderate pressure (10^{-3}-10^{-2} Nm^{-2}), so that when M_2H^+ has a grazing collision with an atom of this gas, a small proportion of its translational (kinetic) energy is converted into internal energy of vibration of M_2H^+. As a result, fragment ions are produced from M_2H^+. In these fragments, the translational energy possessed by the ions M_2H^+ is partitioned in the ratio of the fragment masses. Thus, if a fragment ion retains x% of the mass of M_2H^+, requiring that $(100 - x)$% of the mass is lost as a neutral particle, then the ion retains only x% of the translational energy of M_2H^+. Since the electrostatic analyser deflects ions of lower translational energy more readily, it separates the ions produced by decomposition of M_2^+ according to these energies, and hence in this case according to their masses.

Thus, if the electric sector voltage is scanned downwards from an initial value E_0 (which allows M_2H^+ to pass through the collector slit), the products of the collisionally induced decomposition are detected at the collector in order of decreasing mass. This gives a collisionally induced mass spectrum (CID-MS) of M_2H^+, and these spectra are similar to the mass spectra obtained when the energy is deposited by other means. Since all components of a mixture can be analysed in this way, all components of a mixture M_1, M_2 and M_3 (other than isomers) can in principle be separated, and a mass spectrum obtained for each one. The technique is commonly called MS/MS, since the molecular ion from an initial mass spectrum is selected and made to give a second mass spectrum.

The mass resolution of an electrostatic analyser alone is limited to a few hundred mass units, but greater resolution can be obtained by doing MS/MS with two higher-resolution mass spectrometers in tandem. For example, mixture analysis can be achieved by three quadrupole mass spectrometers connected in sequence. The first quadrupole is used to separate the molecular ions, the second as a collision chamber, and the third to separate the products of collision-induced decomposition. If the mixture contains substances whose structures are novel (i.e. not simply at present unknown to the investigator, but ones that have not previously been determined), then one mass spectrum will not normally be sufficient to determine the structure. The structure determination of such substances will usually require HPLC or GLC (or some other) separation of the components—and subsequent use of a method such as NMR. In contrast, if collision-induced mass spectra are already available and in a computer file, MS/MS can be a sensitive and rapid technique for identifying the components of a mixture.

The Orbitrap mass spectrometer is powerful in carrying out successive MS/MS experiments. In the Orbitrap analyser, all ions except that of a selected molecular ion of interest (say, in a sample which contains several components) may be expelled. By then subjecting this molecular ion to CID, fragments may be obtained from it which help to characterise its structure. But in this case, the convenience lies in being able to target any one of these CID fragments, select its m/z value for retention in the Orbitrap, and then investigate its CID—in the hope of gaining yet more structural information. Such experiments, on whatever instrument they are carried out, are described as $(MS)^n$ experiments, where n is the number of successive CIDs carried out. They can obviously be carried out for each component in a mixture.

4.7 MS data systems

Modern mass spectrometers are controlled using computers. The keyboard is used not only to control the storage, manipulation, and retrieval of data, but, where applicable, also to control the GC or LC. Thus, the flow rate and the temperature gradient in the GC, the solvent proportions in LC, and automatic sampling, injection, and scanning can all be controlled from one site.

The output from a mass spectrometer is a voltage (representing peak intensity) which varies continuously as the spectrum is scanned. This analogue signal is sampled at precise intervals in order to convert it into a digital signal, suitable for handling by the computer. Digital signals of less than a selected amplitude, or arriving in less than a selected time-width, may be rejected in order to remove, respectively, random noise and narrow electrical spikes. The time of arrival of a peak centroid is converted to mass by calibrating the system using the arrival time of peaks of known mass. Some of the processing which is carried out before storage of the digitised spectrum on disc is summarised in Fig. 4.25. The inter-scan report allows the user to monitor spectra by displaying the data on screen as the GC or LC scan progresses, enabling the experiment to be modified as it proceeds.

The computer may also be used to manipulate, and interpret, the data. For example, it is possible to subtract peaks coming from column bleed and from other known impurities. The spectra obtained after such manipulation may then be automatically compared, using the mass and relative abundance of each peak, with a library of mass spectra stored on disk in the computer. This process of library searching for a 'fingerprint' is a powerful method for the identification of compounds whose mass spectra have previously been recorded, and is much used in analytical work.

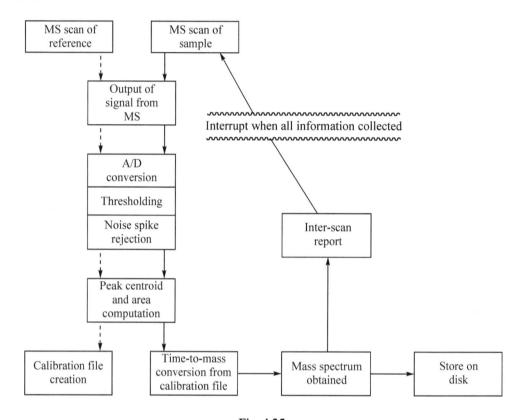

Fig. 4.25

4.8 Specific ion monitoring and quantitative MS (SIM and MIM)

The device normally used to detect ions after mass analysis is an electron multiplier. It operates on the principle that the arriving ions are used to generate electrons, which are then amplified. It is a sensitive device, able at its best to measure a signal from as few as 20 ions. Nevertheless, the scanning of magnetic sector and quadrupole mass spectrometers is inefficient in terms of the signal-to-noise ratio (S/N), because ions are typically recorded for <1% of the time they are produced. If sensitivity is especially important (as, for example, in the analysis of traces of drugs from blood), then S/N can be improved by monitoring only the most abundant ion or ions that are characteristic of the species to be detected. The former process is known as single-ion monitoring (SIM) and the latter as multiple-ion monitoring (MIM). In a magnetic sector instrument, the desired result is achieved by setting the magnet current to scan repetitively through a single-ion signal (SIM); or to scan between a few selected signals (MIM). In the latter case, the signal from each ion is recorded in a different channel of a multi-channel recorder.

The first anti-psychotic drug to be used was chlorpromazine **53**. It was clearly important to know the metabolic profile of this drug in order to be able to assess any physiological effects that the metabolites may possess. The determination of this profile provides an example of MIM in practice. Samples were recovered from the blood of the patient, and converted to their trifluoroacetates (to improve volatility) prior to examination by GC/MS (Fig. 4.26). The mass spectra of chlorpromazine and its side-chain derivatives **54** and **55** contain abundant ions at m/z 246 and 248 (**54**), and 232 and 234 (**55**). Three of these ions, 246, 234 and 232 were monitored continuously from the GC effluent of the derivatised blood extract. The abundance of the three ions rose and fell simultaneously for the two slower eluting compounds, strongly suggesting that these eluants were chlorpromazine derivatives that differed only in the side chains that had been cleaved off. The structures were confirmed by monitoring suspected molecular ions and utilising authentic GC retention times. The two derivatives were, in fact, the trifluoroacetates of the metabolites des- and di-desmethylchlorpromazine. N-Demethylation of drugs containing N-methyl groups is a common metabolic process in the liver.

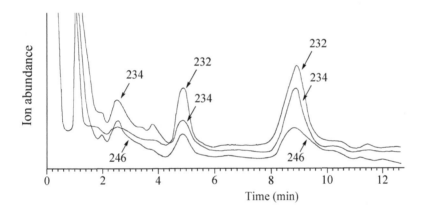

Fig. 4.26

The MIM technique is both remarkably specific and sensitive; in favourable cases, compounds can be identified at the picogram level. The GC/MS/data system combination is powerful in the analysis of pesticide and drug residues, drug metabolites, flavours and perfumes. To make the technique quantitative, a suitable internal standard (e.g. an isotopically labelled analogue) is usually used. For example, a mass spectrometer can readily separate ions produced from the pesticide methyl parathion and a deuterated analogue. If an unknown quantity of methyl parathion is spiked with a known quantity of the deuterated analogue, the relative signal sizes permit a quantitative estimate of the amount of pesticide present.

4.9 Interpreting the spectrum of an unknown

When working on real laboratory problems, it is important to know whether the sample is pure or not. In the case of crystalline compounds, useful information can come from the sharpness, and constancy, of the melting point. More generally, purity can be assessed from thin-layer chromatography (TLC), GC and LC. Characteristic impurity peaks (Table 4.12) may be present in samples that have been extensively handled on thin-layer plates, in columns, or in greased apparatus. Since molecular ions of aliphatic compounds are frequently of low abundance in EI spectra, it is helpful in these cases to determine, in addition, a spectrum using a 'soft' ionisation technique such as CI, FIB or ESI (where a large fraction of the total ion current will be carried by the species characteristic of the molecular weight). For the identification of a peak in a mixture from a chromatographic separation, HPLC/MS or GC/MS, followed by a search of a computer file of EI spectra of known compounds is a powerful method.

In the structure elucidation of an unknown compound, the molecular ion must be identified with certainty. Ensure that the supposed molecular ion is separated from peaks at lower masses by acceptable mass differences (Table 4.13 and Sec. 4.2). Note whether the m/z value of the molecular ion is odd or even, and look for any characteristic isotope patterns. Once a molecular ion, $M^{\cdot+}$, MH^+ or $(M - H^+)^-$, has been identified with certainty, decide whether it is appropriate to determine the molecular formula of the compound by a high-resolution measurement. Be aware that such a measurement will be definitive at low molecular weights but will be consistent with increasing numbers of combinations of elements at higher molecular weights.

After the molecular weight and molecular formula have been determined, use any available UV, IR and NMR spectra and chemical knowledge to determine a suggested partial or total structure. At this stage, return to the mass spectrum to see if the observed fragmentation pattern is consistent with the proposal or limits the range of possibilities, bearing in mind that single-bond cleavages correspond to the energetically more favourable pathways discussed in outline in Sec. 4.5.

In view of the large variety of ion masses that can arise from an equally large variety of organic compounds, the recognition of structural units from m/z values is of limited value, even for molecules of molecular weights <200. It is therefore advantageous to search for the neutrals that are lost from $M^{\cdot+}$, rather than to attempt to identify the fragment ion itself. Table 4.14 lists a number of the common fragments that are easily lost from the molecular ion. Nevertheless, for relatively small molecules where the nature of the functionality is limited and/or known, it is probably worth while to check the observed m/z values for the various fragments with those given in Table 4.15. In doing all this, remember that the real power of mass spectrometry is in the determination of molecular weight and molecular formula, in its sensitivity, and (in the case of a previously isolated compound) the possibility of structure identification through pattern recognition.

In the case of novel molecules, NMR is normally a much more powerful method to determine molecular structure, and mass spectrometry usually acts only in a confirmatory role. On the other hand, mass spectrometry is a major method of structure determination with larger biomolecules, such as polysaccharides, peptides (Table 4.13) and proteins, which are constructed from units of known mass. It is also a major method to facilitate structure determination when there is simply not enough material for a useful NMR spectrum, and in these cases, micro-derivatisation (Table 4.2 on p. 205) is powerful. In these circumstances, as with insect pheromones for example, molecular weight determinations, and fragmentation patterns, which will only be informative if the structures are not too complex, may be all that is available.

4.10 Internet

The Internet is an evolving system, with links and protocols changing frequently. The following information is inevitably incomplete and may no longer apply, but it gives you a guide to what you can expect. Some websites require particular operating systems and may only work with a limited range of browsers, many require payment, and some require you to register and to download programs before you can use them.

For guides to spectroscopic data on the Internet, see the websites at MIT, the University of Waterloo and the University of Texas, representative of several others. They are tailored for internal use, but are informative nevertheless:
http://libraries.mit.edu/guides/subjects/chemistry/spectra_resources.htm
http://lib.uwaterloo.ca/discipline/chem/spectral_data.html
http://www.lib.utexas.edu/chem/info/spectra.html

There are vast resources on the Internet, for MS data. Several of these (CDS, Bio-Rad, ACD, Sigma-Aldrich, ChemGate) cover most or all of the spectroscopic methods, and have already been mentioned on pp. 42-43 in the chapter on IR.

A website listing databases for IR, NMR and MS is:
http://www.lohninger.com/spectroscopy/dball.html

The Japanese Spectral Database for Organic Compounds (SDBS) at:
http://www.aist.go.jp/RIODB/SDBS/cgi-bin/cre_index.cgi
has free access to IR, Raman, ^{1}H- and ^{13}C-NMR and MS data.

The Wiley-VCH website gives access to the SpecInfo data:
http://www3.interscience.wiley.com/cgi-bin/mrwhome/109609148/HOME
and to ChemGate, which has a collection of 700 000 IR, NMR and mass spectra:
http://chemgate.emolecules.com

The Sadtler website is:
http://www.bio-rad.com
follow the leads to Sadtler, KnowItAll and MS for 198 000 mass spectra.
http://www.acdlabs.com/products/spec_lab/exp_spectra/ms/
takes you to the ACD website. In its entry-level configuration, ACD/MS Manager can process MS, tandem mass spectra (MS/MS, MS/MS/MS, MS^n) and MS combined with separation techniques (LC/MS, LC/MS/MS, LC/DAD, CE/MS, GC/MS).

4.11 Bibliography

DEVELOPMENT OF THE SUBJECT

J. H. Beynon, R. A. Saunders and A. E. Williams, *The Mass Spectra of Organic Molecules*, Elsevier, London, 1968.

K. Biemann, *Mass Spectrometry*, McGraw-Hill, New York, 1962.

H. Budzikiewicz, C. Djerassi and D. H. Williams, *Structure Elucidation of Natural Products by Mass Spectrometry*, Vols. I and II, Holden-Day, San Francisco, 1964.

H. Budzikiewicz, C. Djerassi and D. H. Williams, *Mass Spectra of Organic Compounds*, Holden-Day, San Francisco, 1967.

J. R. Chapman, *Computers in Mass Spectrometry*, Academic Press, 1978.

I. Howe, D. H. Williams and R. D. Bowen, *Mass Spectrometry—Principles and Applications,* McGraw-Hill, New York, 1981.

W. H. McFadden, *Techniques of Combined GC/MS*, Wiley, New York, 1973.

B. J. Millard, *Quantitative Mass Spectrometry*, Heyden, London, 1978.

M. E. Rose and R. A. W. Johnstone, *Mass Spectrometry for Chemists and Biochemists*, Cambridge University Press, Cambridge, 1982.

G. R. Waller and O. C. Dermer (Eds), *Biochemical Applications of Mass Spectrometry*, Wiley-Interscience, New York, 1980.

LATER TEXTBOOKS

F. W. McLafferty and F. Turecek, *Interpretation of Mass Spectra*, University Science Books, Sausalito, 4th Ed., 1993.

M. C. McMaster, *GC/MS, A Practical User's Guide*, Wiley, New York, 1998.

C. Herbert and R. Johnstone, *Mass Spectrometry Basics*, CRC Press, Boca Raton, 2002.

E. de Hoffmann and V. Stroobant, *Mass Spectrometry*, Wiley, Chichester, 2nd Ed., 2002.

J. H. Gross, *Mass Spectrometry*, Springer, Heidelberg, 2003.

R. M. Smith, *Understanding Mass Spectra*, Wiley, New York, 2nd Ed., 2004.

N. Nibbering (Ed.), *The Encyclopedia of Mass Spectrometry*, *Vol. 4: Fundamentals of and Applications to Organic (and Organometallic) Compounds*, Elsevier, 2004. This is the most relevant volume of a 10-volume set to structure determination in everyday organic chemistry. The other volumes cover more specialised aspects of the technique.

CHROMATOGRAPHY AND MASS SPECTROMETRY

R. E. Ardrey, *Liquid Chromatography—Mass Spectrometry*, *An Introduction*, Wiley, New York, 2003.

M. C. McMaster, *LC/MS, A Practical User's Guide*, Wiley, New York, 2005.

W. M. A. Niessen, *Liquid Chromatography-Mass Spectrometry*, CRC Press, Boca Raton, 3rd Ed., 2006.

DATA

F.W. McLafferty and D.B. Stauffer, *Wiley/NBS Registry of Mass Spectral Data*, 7 Vols., Wiley, New York, 1989.

F. W. McLafferty and D. B. Stauffer, *Important Peak Index of the Registry of Mass Spectral Data*, 3 Vols., Wiley, New York, 1991.

The Eight Peak Index of Mass Spectra, 3 Vols., 4th Ed., RSC, Cambridge, 1991.

T. J. Bruno and P. D. N. Svoronos, *CRC Handbook of Fundamental Spectroscopic Correlation Charts*, CRC Press, Boca Raton, 2006.

4.12 Tables of data

Table 4.4 Atomic weights and approximate natural abundance of some common isotopes

Isotope	Atomic weight ($^{12}C = 12.000000$)	Natural abundance (%)
^1H	1.007 825	99.985
^2H	2.014 102	0.015
^{12}C	12.000 000	98.9
^{13}C	13.003 354	1.1
^{14}N	14.003 074	99.64
^{15}N	15.000 108	0.36
^{16}O	15.994 915	99.8
^{17}O	16.999 133	0.04
^{18}O	17.999160	0.2
^{19}F	18.998 405	100
^{28}Si	27.976 927	92.2
^{29}Si	28.976 491	4.7
^{30}Si	29.973 761	3.1
^{31}P	30.973 763	100
^{32}S	31.972 074	95.0
^{33}S	32.971 461	0.76
^{34}S	33.967 865	4.2
^{35}Cl	34.968 855	75.8
^{37}Cl	36.965 896	24.2
^{79}Br	78.918 348	50.5
^{81}Br	80.916 344	49.5
^{127}I	126.904 352	100

Table 4.5 ΔH_f (kJ mol^{-1}) of some ions

Ion	ΔH_f	Ion	ΔH_f
H^+	1530	$Me_2C^+\!-\!CH\!=\!CH_2$	770
Me^+	1090	Ph^+	1140
Et^+	920	$PhCH_2^+$	890
$n\text{-}Bu^+$	840	$EtC^+\!=\!O$	600
$EtCH^+Me$	770	$PhC^+\!=\!O$	730
Me_3C^+	690	$MeCH\!=\!OH^+$	600
$CH_2\!=\!CH^+$	1110	$MeO^+\!=\!CH_2$	640
$CH_2\!=\!CH\!-\!CH_2^+$	950	$MeCH\!=\!NH_2^+$	650

Note: 1. Formation of small ions such as H^+ and CH_3^+ is unfavourable. 2. Vinyl and phenyl cations have high ΔH_f values. 3. The ease of formation of alkyl cations is tertiary > secondary > primary. 4. Delocalized cations, and acylium, oxonium, and imminium ions are stabilised in the gas phase, just as they are in solution.

Table 4.6 ΔH_f (kJ mol^{-1}) of some radicals

Radical	ΔH_f	Radical	ΔH_f
H·	218	PhCH$_2$·	188
Me·	142	MeC·=O	−23
Et·	108	HO·	39
n-Pr·	87	MeO·	−4
Me$_2$CH·	74	H$_2$N·	172
CH$_2$=CH·	250	Cl·	122
CH$_2$=CH—CH$_2$·	170	Br·	112
Ph·	300	I·	107

Note the relative instability of vinyl radicals and the increased stability of radicals with increasing substitution.

Table 4.7 Molecular ion abundances in relation to molecular structure

Strong	Medium	Weak or absent
Aromatic hydrocarbons ArH	Conjugated alkenes	Long-chain aliphatics
ArF	Ar⦙Br and Ar⦙I	Branched alkanes
ArCl	ArCO⦙R	Tertiary aliphatics
ArCN	ArCH$_2$⦙R	Tertiary aliphatic bromides
ArNH$_2$	ArCH$_2$⦙Cl	Tertiary aliphatic iodides

Table 4.8 Order of ease of fragmentation of some C$_6$H$_5$X compounds in EI spectra

X	Neutral fragments lost from $M^{\bullet+}$	X	Neutral fragments lost from $M^{\bullet+}$
COMe	Me	OH	CO
CMe$_3$	Me	Me	H
CHMe$_2$	Me	Br	Br
CO$_2$Me	OMe	NO$_2$	NO$_2$ & NO
NMe$_2$	H	NH$_2$	HCN
CHO	H	Cl	Cl
Et	Me	CN	HCN
OMe	CH$_2$O & Me	F	C$_2$H$_2$ & HF
I	I	H	C$_2$H$_2$

Table 4.9 Primary single-bond cleavage processes associated with some common functional groups in an approximate order of ease[†]

Functional group	Fragmentation
Amine	
Acetal	Exceptionally favourable because of stabilised cation
Iodide	
Ether (X=O) Thioether (X=S)	
Ketone	
Alcohol (X=O) Thiol (X=S)	
Bromide	
Ester	

[†]In polyfunctional aliphatic molecules, cleavages associated with functional groups higher up the table are preferred over those cleavages associated with lower entries.

Table 4.10 Useful ion series

Functional group	Simplest ion type	Ion series (m/z)
Amine	$CH_2=NH^+$ m/z 30	30, 44, 58, 72, 86, 100...
Ether and alcohol	$CH_2=OH^+$ m/z 31	31, 45, 59, 73, 87, 101...
Ketone	$MeC\equiv O^+$ m/z 43	43, 57, 71, 85, 99, 113...
Aliphatic hydrocarbon	$C_2H_5^+$ m/z 29	29, 43, 57, 71, 85, 99, 113...

Table 4.11 m/z Values of some McLafferty rearrangement ions found in the mass spectra of carbonyl compounds

Compound	X	m/z
Aldehyde	H	44
Ketone (methyl)	Me	58
Ketone (ethyl)	Et	72
Acid	OH	60
Ester (methyl)	OMe	74
Ester (ethyl)	OEt	88
Amide	NH_2	59

Table 4.12 Some common impurity peaks

m/z Values	Cause
149, 167, 279	Plasticizers (phthalic acid derivatives)
129, 185, 259,329	Plasticizer (tri-n-butyl acetyl citrate)
133, 207, 281, 355, 429	Silicone grease
99, 155, 211	Plasticizer (tributyl phosphate)

Table 4.13 Integral masses of amino acid residues —NHC(R)CO— with the one-letter code in parenthesis after the three-letter code

Gly(G)	57	Ser(S)	87	Gln(Q)	128	His(H)	137
Ala(A)	71	Thr(T)	101	Phe(F)	147	Pro(P)	97
Val(V)	99	Asp(D)	115	Tyr(Y)	163	Met(M)	131
Leu(L)	113	Asn(N)	114	Trp(W)	186	Arg(R)	156
Ile(I)	113	Glu(E)	129	Lys(K)	128	Cys(C)	103

Based on C = 12.0, H = 1.0, N = 14.0 and O = 16.0. When the first amino acid is lost from the carboxyl terminus of a peptide to give an ion of type **49**, the mass lost from $(M - H)^-$ or MH^+ is one mass unit greater than the values given in the table.

Table 4.14 Some common losses from molecular ions

Ion	Groups commonly associated with the mass lost	Possible inference
M − 1	H	
M − 2	H_2	
M − 14		Homologue?
M − 15	CH_3	
M − 16	O	$ArNO_2$. $R_3N{-}O$, R_2SO
M − 16	NH_2	$ArSO_2NH_2$, $-CONH_2$
M − 17	OH	
M − 17	NH_3	
M − 18	H_2O	Alcohol, aldehyde, ketone, etc.
M − 19	F	Fluoride
M − 20	HF	Fluoride
M − 26	C_2H_2	Aromatic hydrocarbon
M − 27	HCN	Aromatic nitrile, nitrogen heterocycle
M − 28	CO	Quinone
M − 28	C_2H_4	Aromatic ethyl ether, ethyl ester, n-propyl ketone
M − 29	CHO	
M − 29	C_2H_5	Ethyl ketone, n-Pr—Ar
M − 30	C_2H_6	
M − 30	CH_2O	Aromatic methyl ether
M − 30	NO	Ar—NO_2
M − 31	OCH_3	Methyl ester
M − 32	CH_3OH	Methyl ester
M − 32	S	
M − 33	$H_2O + CH_3$	
M − 33	SH	Thiol
M − 34	H_2S	Thiol
M − 41	C_3H_5	Propyl ester
M − 42	CH_2CO	Methyl ketone, aromatic acetate, $ArNHCOCH_3$,
M − 42	C_3H_6	n- or i-butyl ketone, aromatic propyl ether, n-Bu—Ar
M − 43	C_3H_7	Propyl ketone, n-Pr—Ar
M − 43	CH_3CO	Methyl ketone
M − 44	CO_2	Ester (skeleton rearrangement, anhydride)
M − 44	C_3H_8	
M − 45	CO_2H	Carboxylic acid
M − 45	OC_2H_5	Ethyl ester
M − 46	C_2H_5OH	Ethyl ester
M − 46	NO_2	Ar—NO_2
M − 48	SO	Aromatic sulfoxide
M − 55	C_4H_7	Butyl ester
M − 56	C_4H_8	n- or i-C_5H_{11}—Ar, n- or i-Bu—OAr, pentyl ketone
M − 57	C_4H_9	Butyl ketone
M − 57	C_2H_5CO	Ethyl ketone
M − 58	C_4H_{10}	
M − 60	CH_3CO_2H	Acetate

Table 4.15 Masses and some possible compositions of common fragment ions

m/z	Groups commonly associated with the mass	Possible inference
15	CH_3^+	
18	H_2O^+	
26	$C_2H_2^+$	
27	$C_2H_3^+$	
28	$C_2H_3^+CO^+, C_2H_4^+, N_2^+$	
29	$CHO^+, C_2H_5^+$	
30	$CH_2=NH_2^+$	Some primary amines
31	$CH_2=OH^+$	Some primary alcohols
36/38 (3:1)	HCl^+	
39	$C_3H_3^+$	
40	$Argon^{+a}, C_3H_4^+$	
41	$C_3H_5^+$	
42	$C_2H_2O^+, C_3H_6^+$	
43	CH_3CO^+	CH_3CO—X
44	$C_2H_6N^+$	Some aliphatic amines
44	$O=C=NH_2^+$	Primary amides
44	$CO_2^+, C_3H_8^+$	
44	$CH_2=CH(OH)^+$	Some aldehydes
45	$CH_2=O^+CH_3, CH_3CH=OH^+$	Some ethers and alcohols
47	$CH_2=SH^+$	Aliphatic thiol
49/51 (3:1)	CH_2Cl^+	
50	$C_4H_2^+$	Aromatic compound
51	$C_4H_3^+$	C_6H_5—X
55	$C_4H_7^+$	
56	$C_4H_8^+$	
57	$C_4H_9^+$	C_4H_9—X
57	$C_2H_5CO^+$	Ethyl ketone, propionate ester
58	$CH_2=C(OH)CH_3^+$	Some methyl ketones, some dialkyl ketones
58	$C_3H_8N^+$	Some aliphatic amines
59	$CO_2CH_3^+$	Methyl ester
59	$CH_2=C(OH)NH_2^+$	Some primary amides
59	$C_2H_5CH=OH^+$	$C_2H_5CH(OH)$—X
59	$CH_2=O^+-C_2H_5$ and isomers	Some ethers

[a]The argon doublet from the air is a useful reference in counting mass peaks.

Table 4.15 *continued*

m/z	Groups commonly associated with the mass	Possible inference
60	CH$_2$=C(OH)OH$^+$	Some carboxylic acids
61	CH$_3$CO(OH$_2$)$^+$	CH$_3$CO$_2$C$_n$H$_{2n+1}$ (n>1)
61	CH$_2$CH$_2$SH$^+$	Aliphatic thiol
66	H$_2$S$_2$$^+$	Dialkyl disulfide
68	CH$_2$CH$_2$CH$_2$CN$^+$	
69	CF$_3$$^+$, C$_5H_9$$^+$	
70	C$_5$H$_{10}$$^+$	
71	C$_5$H$_{11}$$^+$	C$_5$H$_{11}$—X
71	C$_3$H$_7$CO$^+$	Propyl ketone, butyrate ester
72	CH$_2$=C(OH)C$_2$H$_5$$^+$	Some ethyl alkyl ketones
72	C$_3$H$_7$CH=NH$_2$$^+$ and isomers	Some amines
73	C$_4$H$_9$O$^+$	
73	CO$_2$C$_2$H$_5$$^+$	Ethyl ester
73	(CH$_3$)$_3$Si$^+$	(CH$_3$)$_3$Si—X
74	CH$_2$=C(OH)OCH$_3$$^+$	Some methyl esters
75	C$_2$H$_5$CO(OH$_2$)$^+$	C$_2$H$_5$CO$_2$C$_n$H$_{2n+1}$ (n>1)
75	(CH$_3$)$_2$Si=OH$^+$	(CH$_3$)$_3$SiO—X
76	C$_6$H$_4$$^+$	C$_6$H$_5$—X, X—C$_6$H$_4$—Y
77	C$_6$H$_5$$^+$	
78	C$_6$H$_6$$^+$	
79	C$_6$H$_7$$^+$	
79/81 (1:1)	Br$^+$	
80/82 (1:1)	HBr$^+$	
80	C$_5$H$_6$N$^+$	
81	C$_5$H$_5$O$^+$	
83/85/87 (9:6:1)	HCCl$_3$$^+$	CHCl$_3$
85	C$_6$H$_{13}$$^+$	C$_6$H$_{13}$—X
85	C$_4$H$_9$CO$^+$	C$_4$H$_9$CO—X

Table 4.15 *continued*

m/z	Groups commonly associated with the mass	Possible inference
85	[dihydropyran oxocarbenium structure]	[tetrahydropyran–X structure]
85	[butyrolactone cation structure]	[X–butyrolactone structure]
86	$CH_2=C(OH)C_3H_7^+$	Some propyl alkyl ketones
86	$C_4H_9CH=NH_2$ & isomers	Some amines
87	$CH_2=CHC(=OH^+)OCH_3$	$X—CH_2CH_2CO_2CH_3$
91	$C_7H_7^+$	$C_6H_5CH_2—X$
92	$C_7H_8^+$	$C_6H_5CH_2—alkyl$
92	$C_6H_6N^+$	[pyridine–CH2X structure]
91/93 (3:1)	[cyclopentyl chloride cation structure]	n-Alkyl chloride ($\geq$hexyl)
93/95 (1:1)	CH_2Br^+	
94	$C_6H_6O^+$	$C_6H_5O—alkyl (alkyl > CH_3)$
94	[pyrrole acylium structure]	[pyrrole–C(=O)X structure]
95	[furan acylium structure]	[furan–C(=O)X structure]
95	$C_6H_7O^+$	[Me–furan–C(=O)X structure]
97	$C_5H_5S^+$	[thiophene–CH2X structure]

Table 4.15 *continued*

m/z	Groups commonly associated with the mass	Possible inference
99		
99		
105	$C_6H_5CO^+$	C_6H_5CO—X
105	$C_8H_9^+$	$CH_3C_6H_4CH_2$—X
106	$C_7H_8N^+$	
107	$C_7H_7O^+$	
107/109 (1:1)	$C_2H_4Br^+$	
111		
121	$C_8H_9O^+$	
122	$C_6H_5CO_2H^+$	Alkyl benzoates
123	$C_6H_5CO_2H_2^+$	Alkyl benzoates
127	I^+	
128	HI^+	
135/137 (1:1)		n-Alkyl bromide (≥hexyl)
130	$C_9H_8N^+$	
141	CH_2I^+	
147	$(CH_3)_2Si=O$—$Si(CH_3)_3$	

Table 4.15 *continued*

m/z	Groups commonly associated with the mass	Possible inference
149		Dialkyl phthalate
160	$C_{10}H_{10}NO^+$	
190	$C_{11}H_{12}NO_2^+$	

5. Practice in structure determination

5.1 General approach

There is no fixed way of tackling a problem in structure determination using the four spectroscopic methods. Each problem has its own unique features, and some knowledge of the provenance makes a big difference to how one starts. At one extreme, the product of a reaction with known starting materials is often predictable, and the purpose of the spectroscopic investigation is largely to check that the compound actually isolated does have the expected structure. On the other hand, a complete unknown, extracted from a plant for example, will have little information to be drawn on. In the former case one must guard against a too easy assumption that the predicted structure fits the spectra; in the latter one can only draw on one's knowledge of the known structures of natural products—often of limited use since these are astonishingly diverse. In the discussion that follows, we work through some representative examples in which provenance is missing. This is not all that realistic, since compounds handled in research rarely have so little information attached to them, but these examples will serve to show how one can go from one spectrum to the next to draw out all the information needed in order to assign a structure. The order in which we put together the pieces of spectroscopic information in order to deduce the structures here is by no means the only one that would work, or even necessarily the shortest. We hope that what we do here serves to show how expeditiously the various leads can be connected.

It is almost always best to begin with the molecular weight of the unknown derived from its mass spectrum, and ideally to obtain the molecular formula from a high-resolution measurement. In the absence of a high-resolution measurement, combustion analysis may afford the same information; but a note of caution is necessary. Deductions based upon combustion analysis are prone to errors, either from random experimental error (making at least two consistent determinations desirable), or from the presence of impurities, although the power of modern HPLC (High Performance Liquid Chromatography) has reduced the latter problem. Except in the case of molecules made up from standard molecular building blocks (e.g. peptides), mass spectral fragmentation is of lesser importance—the information it provides is crucial only in exceptional cases, and is usually looked at later in the analysis. The spectra of greatest power in structure determination are ^{13}C and ^{1}H NMR, and UV and IR are less important. Sometimes, UV provides critical information on the extent of conjugation, and IR on the presence of a functional group, or, perhaps critically, the size of ring in which a carbonyl group is incorporated, but otherwise these methods are apt to take a back seat.

If the mass spectrum has given the molecular formula, it is wise to work out the number of 'double-bond equivalents' (DBE) in a molecule. This is done by inspecting the molecular formula. If the molecule contains no elements other than C, H, N and O, then the number of DBE is given by Eq. 5.1.

$$C_aH_bN_cO_d \qquad\qquad DBE = \frac{(2a+2)-(b-c)}{2} \qquad\qquad (5.1)$$

The $(2a+2)$ term is the number of hydrogens in a saturated hydrocarbon having a carbon atoms. Relative to such a structure, every ring or double bond necessitates two fewer hydrogen atoms (cyclohexane is C_6H_{12}, and so is 1-hexene). Therefore, subtracting b, the actual number of hydrogens present, from $(2a+2)$ and dividing by two gives the total number of double bonds and rings in the molecule. It is useful to remember that a benzene ring has a total of four double-bond equivalents: three 'double bonds' and one ring. The number of divalent atoms present makes no difference to this calculation, but the number of mono- and trivalent atoms does. Count monovalent atoms (F, Cl, Br, etc.) as hydrogens and add them to b. When c trivalent atoms (N, trivalent P, etc.) are present, subtract c from b for each trivalent atom present. Thus, the formula $C_5H_{11}N$ corresponds to one double-bond equivalent (often written: $\boxed{1}$); it might, for example, be the N,N-dimethyl enamine of propanal ($Me_2NCH{=}CHMe$), with one double bond, or cyclopentylamine, with one ring. Many other structures are possible, but we can already see that, if one of the spectroscopic methods reveals the presence of a double bond ($C{=}C$, $C{=}O$ or $C{=}N$), then there are no rings. Conversely, if double bonds clearly are not present, then there must be a ring.

It is sensible to look next at the ^{13}C and ^1H spectra, with a quick glance at the IR to pick up the presence, or absence, of a distinctive functional group. For this reason, in Secs. 5.2 and 5.3 we give examples in each of which the interpretation of the ^{13}C and ^1H spectra alone suffices, and in Sec. 5.4 we work more thoroughly through six examples in which the four spectroscopic techniques are used in combination to derive the structures of slightly more complicated molecules.

5.2 Simple worked examples using ^{13}C NMR alone

The ^{13}C NMR spectrum gives a wealth of information, often making it the first place to look after a formula has been deduced. It gives the number of differently situated carbon atoms in the molecule, easily subdivided by chemical shift into digonal, trigonal and tetrahedral carbons. It gives the numbers of fully substituted, CH, CH_2 and CH_3 carbons, and it gives information about symmetry elements, which cause two or more carbons to be in identical environments. Furthermore, some functional groups, such as carbonyl groups, have distinctive signals. Since proton-decoupled carbon spectra are simple, the information they contain is often presented in numerical form, simply as a list of chemical shifts, with each shift being followed by (s), (d), (t) or (q), which indicate the $^1J_{CH}$ coupling which would give a singlet, doublet, triplet or quartet, respectively, if the spectrum were to have been taken without decoupling. This information identifies the fully substituted, CH, CH_2 and CH_3 carbons, and will have been derived from an off-resonance decoupled spectrum (Sec. 3.5) or, more likely today, from an APT or DEPT spectrum (Sec. 3.15). Longer-range C—H coupling, $^2J_{CH}$ and $^3J_{CH}$, although examined at length in Sec. 3.5, is rarely used at this stage, and when it is it is taken from 2D HMQC and HMBC spectra (Sec. 3.21), rather than from the 1D ^{13}C spectrum.

In tackling problems in which the information is presented in digital form, you may find it helpful to draw out an approximate analogue spectrum, and to consider it as divided into three regions, starting at the low-field end on the left:

1. The region from 220 to 160 p.p.m., in which the various kinds of carbonyl carbons will occur. Remember as a reference point that the carbonyl carbons of aldehydes and ketones occur in the range 220-190 p.p.m. (Table 3.13). Aldehydes can be distinguished easily from ketones, since the former give rise to doublets. Since electron donation to the carbonyl carbon shields this atom, it can be expected that carboxylic acids, esters and amides will resonate at higher field, that is lower p.p.m. values—in the 190-160 p.p.m. range (Table 3.13).

2. The region from 160 to 100 p.p.m., in which the various kinds of trigonal carbons other than carbonyl carbons resonate.

3. The region from 100 to 0 p.p.m., in which the various kinds of tetrahedral carbons resonate. Note that although electronegative substituents bring tetrahedral carbons towards the high values in this range, even the anomeric carbons of sugars, with two attached oxygen atoms, are deshielded to only *ca.* 100 p.p.m. By counting the carbonyl and trigonal carbons, the numbers of double bonds in the unknown can be estimated, but bear in mind that nitrile carbons (—C≡N) occur in the 120-105 p.p.m. region and acetylene carbons (—C≡C—) in the 90-65 p.p.m. region.

Example 1

Two isomeric hydrocarbons C_5H_{10}, **1** and **2**, were separated by GC and gave the following ^{13}C data. **1**: δ 132(s), 118(d), 26(q), 17(q) and δ 13(q). **2**: δ 147(s), 108(t), 31(t), 22(q) and 13(q). Deduce structures for the olefins.

The hydrocarbons contain one DBE, which is present as a single double bond (each gives two signals in the 100-160 p.p.m. range). Each has no equivalent carbons, and the multiplicities indicate the 10 required hydrogens: in **1** as three methyl groups and one trigonal CH, and in **2** as two methyl groups, one tetrahedral CH_2, and one trigonal CH_2. The structures are therefore:

1   2

Example 2

A compound gives a molecular ion in its EI spectrum at *m/z* 94, unchanged when the compound is introduced into the mass spectrometer in the presence of D_2O. A high-resolution measurement establishes the molecular formula as $C_5H_6N_2$. The compound gives signals in its ^{13}C NMR spectrum at δ 119(s), 22(t) and 16(t). Deduce its structure.

The compound has four DBEs. Since the molecular weight is unchanged in the presence of D_2O, the six hydrogens must be attached to carbon. The carbon atoms giving rise to the triplet signals at δ 22 and 16 carry two protons each, and therefore, to give six hydrogens in total, one of these signals must correspond to two carbons. Hence, to give five carbons in total, the signal at 119 must also correspond to two carbons, neither of which has any hydrogen atoms attached to it.

If the signal at 119 were from two trigonal carbons, then we could account for only one DBE in this way—clearly not enough. Also, acetylenic carbons (δ 90-65) are precluded, but the two carbons at 119 are consistent with the presence of two nitriles (Table 3.9), which does account for four DBE. This leaves the signals at 16 and 22 to account for three tetrahedral carbons in an acyclic structure **3**, which also accommodates the symmetry required by the data.

3

5.3 Simple worked examples using ^{1}H NMR alone

Simple ^{1}H NMR spectra can be represented in digital form, just like ^{13}C spectra. Singlets, doublets, triplets and quartets are reported as (s), (d), (t) and (q), quintets and sextets (and beyond) are spelled out to avoid ambiguity, unresolved or uninterpretable multiplet signals are indicated by (m), and broad signals by (br). It is again not uncommon to work from left to right, from low to high field. Given the molecular formula of an unknown, the nature of the functional groups can often be inferred, and even complete structures derived solely from these spectra.

Examples 3-5

Three isomeric compounds $C_4H_8O_3$, each of which must therefore contain one DBE, give the following ^{1}H NMR spectra. Deduce plausible structures for each of them.

Example 3: δ 12.1 (1H, s), 4.15 (2H, s), 3.6 (2H, q, J = 7 Hz) and 1.3 (3H, t, J = 7 Hz). The quartet signal at δ 3.6 in the ^{1}H NMR spectrum immediately suggests the presence of an OCH_2CH$_3$ group—its position of resonance is appropriate for an OCH$_2$ group (Table 3.19) and the multiplicity tells us that a methyl group is joined to it. The OCH$_2$CH_3 signal is also present as a triplet at δ 1.3. In more detail, the precise chemical shift of the lower field signal—just a little to higher field of 4 p.p.m.—suggests an ethyl ether (OC$_2$H$_5$) rather than an ethyl ester (RCOOC$_2$H$_5$) (Table 3.19). The very low-field proton at δ 12.1 is consistent with the presence of a carboxylic acid (Table 3.27), plausibly accounting for the one DBE. Thus, all that remains is a CH$_2$ group, which must be placed so as not to be coupled, and its chemical shift at δ 4.15 indicates that it is adjacent to an electronegative element. Therefore the structure is CH$_3$CH$_2$OCH$_2$COOH.

Example 4: δ 4.15 (1H, 1:5:10:10:5:1 sextet, J = 7 Hz), 2.35 (2H, d, J = 7 Hz) and 1.2 (3H, d, J = 7 Hz). Spectrum run in D$_2$O.
Only six of the eight protons are observed in the spectrum, indicating that two protons are probably bound to oxygen, and therefore after exchange with D$_2$O become OD groups. The integrated areas of the signals, in conjunction with their multiplicities, indicate the structural element CH$_3$CHCH$_2$, to which the atoms CH$_2$O$_3$ must be attached. The observed deuterium exchange behaviour and the water solubility of the compound can therefore be satisfied by OH and CO$_2$H groups so that two structures might fit: CH$_3$CH(OH)CH$_2$CO$_2$H or CH$_3$CH(CO$_2$H)CH$_2$OH. The methine proton observed in the spectrum (δ 4.15) is in good accord with the expected chemical shift (δ 3.9 ± 0.2 in Table 3.19) if the former structure is correct, with the methine proton's being β to the carboxyl group taking it just downfield of the normal range.

Example 5: δ 4.05 (2H, s), 3.8 (3H, s) and 3.5 (3H, s).

The two three-proton singlets suggest the presence of two methoxyl groups. Perhaps one of these is part of a methyl ester (δ 3.8) and the other part of a methyl ether (δ 3.5), a proposal which takes care of the three oxygen atoms, leaving only a CH_2 group to satisfy the molecular formula. The two-proton uncoupled signal at δ 4.05 must come from that methylene group, which is downfield because it is next to an ether oxygen atom and next to a carbonyl group. The data are therefore consistent with the structure $CH_3OCH_2CO_2CH_3$.

5.4 Simple worked examples using the combined application of all four spectroscopic methods

In the following examples, we often incorporate UV and IR information into the structure elucidation. In so doing, we probably give undue emphasis to these methods relative to ^{13}C and ^1H NMR, but the benefit is to illustrate how UV and IR fit into the wider scheme. See how much of the structure you can deduce for yourself from the spectra, and only then read our analysis.

Example 6

The spectra for Example 6 are on the next page. The mass spectrum and the combustion analysis show that the formula is C_4H_8O, and there is therefore one double-bond equivalent. The UV spectrum has λ_{max} 295 nm, but before doing anything with this information, it is important always to work out the ε value from Eq. 1.2 in order to find out how intense the absorption is. Eq. 1.2 is usefully rewritten as Eq. 5.2.

$$\varepsilon = \frac{\text{absorbance} \times \text{molecular weight} \times 100}{\text{weight of compound in mg in 100 ml} \times \text{path length in cm}} \tag{5.2}$$

In this case, the numbers are:

$$\varepsilon = \frac{0.28 \times 72 \times 100}{106 \times 1} = 19$$

The absorption is therefore very weak, and typical of the n$\rightarrow\pi^*$ absorption of a saturated ketone or aldehyde (Sec. 1.15). One should not place too much faith in this information alone—the absorption could so easily be a trace of strongly absorbing impurity, which the other spectroscopic methods would not detect. However, in this case, the presence of a carbonyl group is immediately apparent in the IR spectrum with its very strong band at 1715 cm^{-1}; furthermore, it is clearly a ketone and not an aldehyde, since the latter would have a carbonyl band at slightly higher frequency (Table 2.7) together with further absorption between 2900 and 2700 cm^{-1} (Table 2.1), and, most tellingly, absorption in the δ 9-10 region of the ^1H NMR spectrum. With only one double bond equivalent and one heteroatom to account for, we have now found them both in a ketone carbonyl group. There can be no further functionality, so all we have to do is arrange the carbon skeleton.

Here we could look equally profitably at either the NMR or mass spectrum. The ^{13}C NMR shows that all four carbon atoms are different; one (at δ 208.8) is clearly the carbonyl carbon, being both weak and at a very low field (Table 3.13). As it happens, in this very simple case, we already have all the information we need with which to deduce a structure, but typically we would look next at the ^1H NMR spectrum, where we see a quartet (δ 2.44), a singlet (δ 2.13) and a triplet (δ 1.04). The integration trace on a large print-out gives a rise of 114 mm for the quartet, 174 mm for the singlet, and 175 mm for

Spectra for example 6

C, 66.7%; H, 11.1%

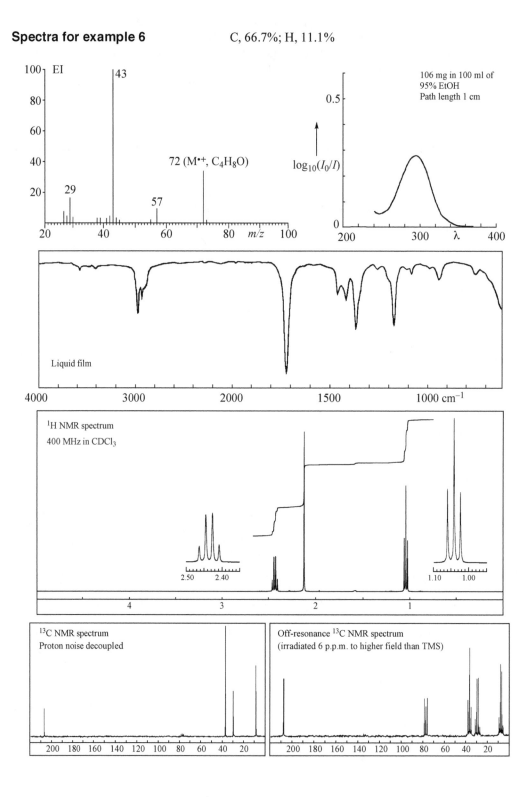

the triplet. Since there is a total of eight hydrogen atoms in the molecule, these correspond to 1.97H, 3.01H and 3.02H, respectively; clearly the true ratios are 2:3:3. The three-proton signal at δ 2.13 will be a methyl group; since it is a singlet, it must be attached to an atom having no protons on it. Its position of resonance is compatible (Table 3.19) with its being a $CH_3CO—$ group. The 1:3:3:1 quartet and the 1:2:1 triplet integrating for two and three protons, respectively, are typical of an ethyl group, as we have already seen in Example 3. Since the methylene resonance is in the δ 1.4-2.5 range (Table 3.19, column 2), the methylene group this time must be attached to a carbon atom, and, since it only gives rise to a quartet, it must be attached to an atom having no hydrogen atoms on it. Clearly it is joined directly to the carbonyl carbon, and the whole structure is, $CH_3COCH_2CH_3$.

Although we have unambiguously solved the structure at this stage, it is always wise in the general case to look at the other spectra and confirm that we have not been misled. The positions of the ^{13}C signals, for instance, are obviously right for this structure: the peak at 7.86 is the methyl group in the ethyl group, $CH_3COCH_2CH_3$, the peak at 29.37 is the methyl group next to the ketone, $CH_3COCH_2CH_3$, and the peak at 36.80 is the methylene group, $CH_3COCH_2CH_3$ (Table 3.6). These assignments are confirmed by the off-resonance spectrum (Sec. 3.5), in which the same signals are a quartet, a quartet and a triplet, showing that the corresponding carbon atoms carry three, three and two hydrogens, respectively. Note that the carbonyl signal is still a singlet in the off-resonance spectrum.

The mass spectrum, too, is definitive: the molecule breaks at each of the C—CO bonds **4** (Table 4.9), and hence detects the methyl and ethyl groups directly.

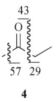

4

Example 7

The spectra for Example 7 are on the next page. The molecular formula $C_{11}H_{20}O_4$ shows that there are two double-bond equivalents, and the absence of UV absorption shows that these are not from two conjugated double bonds. The IR spectrum shows a strong carbonyl band at 1740 cm^{-1}, which could be a five-ring ketone or a saturated ester. There is no C=C double-bond absorption in the IR, so the two double-bond equivalents are either two carbonyl groups or one carbonyl group and a ring. There is no OH absorption in the IR, so the four oxygen atoms must be in ketone, ester or ether groups.

The ^{13}C NMR spectrum is informative at this stage in the analysis—it shows only eight different kinds of carbon atom in a molecule having 11 in all (we discount the three small peaks from the CDCl$_3$ solvent). Some carbon atoms must be repeated in identical structures symmetrically disposed in the molecule. The 1:3:3:1 quartet at δ 4.2, and the 1:2:1 triplet at δ 1.27 in the ^{1}H NMR spectrum suggest the presence of an OCH$_2$CH$_3$ group, as we saw in Example 3. In more detail, the precise position of the OCH$_2$ resonance (δ 4.2) suggests (Table 3.19) that the OCH$_2$CH$_3$ group is actually an ester, $CO_2CH_2CH_3$, and not an ether. The intensity of the OCH$_2$ signal corresponds to *four*

Spectra for example 7   C, 61.0%; H, 9.4%

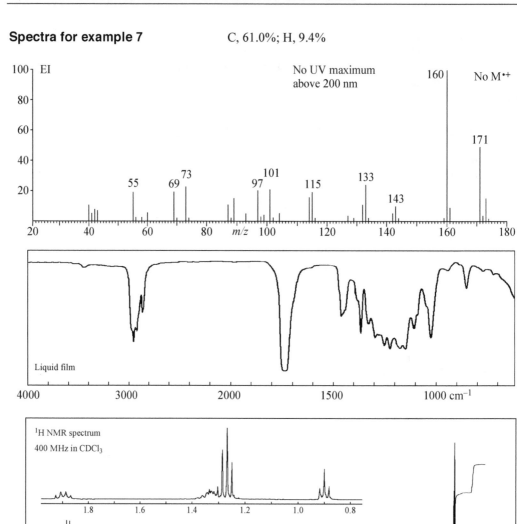

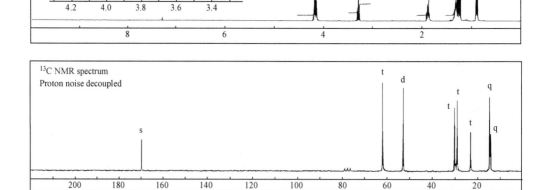

hydrogens, which means that there are two (possibly identical) $CO_2CH_2CH_3$ groups, thus accounting for the presence of only eight different carbon atoms in the molecule, with three of them duplicated in the CO_2Et groups.

The next informative signal to look at is the triplet at δ 3.3: integration makes this a single hydrogen, and we have to account for the chemical shift at which it resonates. Since there is no other functionality in the molecule (the CO_2Et groups account for the two double-bond equivalents and all the heteroatoms), a one-proton triplet at this position can only be produced by the grouping —$CH_2CH(CO_2Et)_2$, the carbethoxy groups causing the downfield shift of the hydrogen near them. Furthermore, the two-proton quartet at δ 1.9 is likely to be the signal from the —$CH_2CH(CO_2Et)_2$ hydrogens, and, since it is a quartet, it too is bonded to a CH_2 group (to make the total number of vicinal hydrogens three). Thus, we have the fragment —$CH_2CH_2CH(CO_2Et)_2$. The three-proton triplet at δ 0.9 can only be produced by a CH_3CH_2 group; the CH_2 signal of this group, and of the —$CH_2CH_2CH(CO_2Et)_2$ group, give the unresolved multiplet at δ 1.33. The two fragments we have now identified, CH_3CH_2— and —$CH_2CH_2CH(CO_2Et)_2$, account for all the atoms of the molecular formula, and the structure must be made by joining them together **5**.

This structure also fits the ^{13}C NMR spectrum: 1. There are two different kinds of methyl group (at δ 14.10 and 13.81), with the signal at a lower field about twice as intense (because it comes from two identical methyl groups). We know that they are methyl groups, not only because of their chemical shift but also because they are displayed on the opposite side of the base line from the $CDCl_3$ signal in the APT spectrum (Sec. 3.15). To save space, the DEPT, APT or off-resonance spectrum is not shown in this or any of the spectra from now on, but the deduced multiplicity is shown on the completely decoupled spectra as a letter (s, d, t and q). 2. There are three C—CH_2—C groups (labelled t, appearing on the same side of the base line as the $CDCl_3$ signal in the APT spectrum) at δ 29.5, 28.5 and 22.4. We discussed the assignment of these signals in Sec. 3.29 in the example of the application of Eq. 3.16. 3. There is a methine group (labelled d) at δ 52.0. 4. There is the CH_2O carbon at δ 61.1 and the carbonyl carbon at δ 169.3.

The mass spectrum also confirms the structure 2, and the base peak, unusual in being an even number (m/z 160), is the result of β-cleavage with γ-hydrogen rearrangement **6** (Sec. 4.5 and Table 4.11).

Example 8

The mass spectrum has no molecular ion, as we can tell by the presence of a large number of peaks immediately to lower molecular mass than the strong peak at m/z 55. The combustion analysis gives a formula, $C_5H_{11}NO_4$. There is therefore one double-bond equivalent, which is clearly not a carbonyl group because there is no absorption in the IR spectrum between 1900 and 1600 cm^{-1} and no signal in the ^{13}C NMR at appropriately low field. Instead, there is very strong absorption at 1545 cm^{-1}, which is likely to be from a nitro group (Table 2.11), a formulation supported in the first place by the UV spectrum with a weak n→π* band at 275 nm (ε 24), and in the second by the absence of the molecular ion in the mass spectrum, which is common with aliphatic nitro compounds, because of the ease with which the NO$_2$ radical is released. The IR spectrum, with strong absorption at 3350 cm^{-1}, also shows the unmistakable presence of a hydroxyl group.

The ^{13}C NMR spectrum shows that there are only four different kinds of carbon atom, which means that two carbons are in the same magnetic environment and are most likely to be two identical groups symmetrically disposed in the molecule. The next most useful piece of information is from the ^{1}H NMR spectrum. The integration distributes the 11 hydrogens thus: two hydrogens in each of the signals at δ 4.2 and 3.9, two hydrogens to the multiplet (actually an ill-resolved triplet) at δ 2.95, two hydrogens to the quartet at δ 1.92, and three hydrogens to the triplet at δ 0.92. These last two signals (the two-proton quartet and the three-proton triplet) are obviously an ethyl group, and the chemical shift of the quartet means it must be bound to a carbon atom. The fact that it is only a quartet means that the carbon atom is fully substituted. Going back to the ^{13}C NMR spectrum, the C-ethyl group is responsible for the quartet at δ 7.66 (—CH$_2$CH$_3$) and either the triplet at δ 63.51 or the triplet at δ 25.77 (—CH$_2$CH$_3$). The fully substituted carbon atom must be the weak singlet at δ 94.23 (fully substituted carbons are usually weak, see Sec. 3.3). Thus, the two identical carbon atoms must be responsible for whichever of the two triplets (δ 63.51 and 25.77) is not produced by the CH$_2$ group of the ethyl group. One of these (δ 63.51) is nearly twice as intense as the other, so it is likely that it comes from the two identical groups and the other (δ 25.77) from the ethyl group. The two identical groups are therefore CH$_2$ groups (labelled t, because they are on the same side of the base line as the CDCl$_3$ signal in the APT spectrum), and they are at a comparatively low field, suggesting that perhaps they are attached to an electronegative heteroatom (Table 3.6).

When the ^{1}H NMR spectrum is taken after a shake with D$_2$O (Sec. 3.4.2), the two-proton multiplet at δ 2.95 disappears, showing that there are two OH groups. At the same time the doubling of the doublets at δ 4.2 and 3.9 also disappears, showing that the OH groups are attached to CH$_2$ groups in which the hydrogen atoms are not equivalent. It is now clear that the two identical groups are CH$_2$OH groups, and we have found the fragments **7**, **8** and **9**. These account for all the atoms, and there is only one way of putting them together, namely **10**.

$$—NO_2 \qquad 7$$

$$8 \quad \begin{array}{c} X \\ Z \ \ Y \end{array}$$

$$9 \quad (\ \ OH)_2$$

$$10 \quad HO \begin{array}{c} NO_2 \\ \\ OH \end{array}$$

A remarkable feature of the ^{1}H NMR spectrum still remains to be examined—the complex signal from the CH$_2$OH hydrogens. After the D$_2$O shake this is an AB quartet (δ_A 4.21, δ_B 3.98, J_{AB} = 12 Hz). The CH$_2$OH groups are bonded to a prochiral centre, not

Spectra for example 8   C, 40.2%; N, 9.5%; H, 7.3%

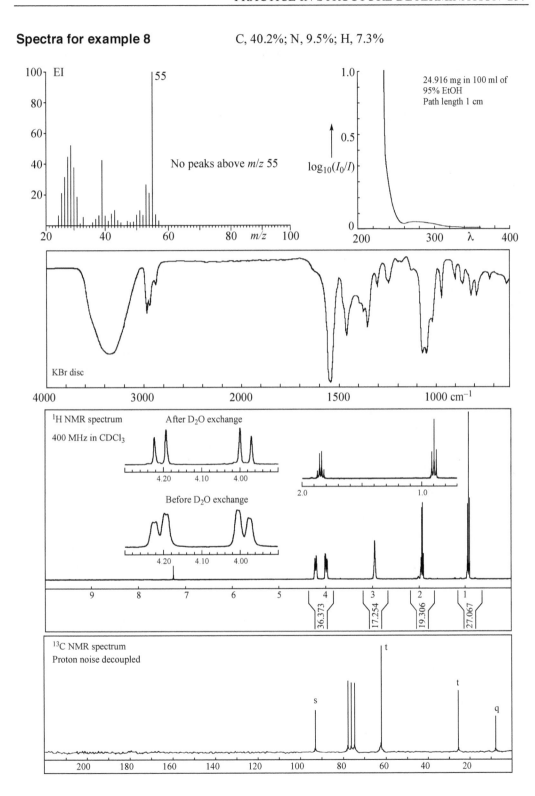

a stereogenic centre, but the A and B hydrogens of each of the —CH_AH_BOH groups 'see' three different groups on the adjacent carbon atom. They do not necessarily experience the same magnetic environment and can, as in this case, resonate with different chemical shifts, and couple with each other in a similar way to the signal from the diastereotopic methylene groups in Fig. 3.32, but in this case giving rise to a simple AB system. Before the D_2O shake, they also couple with the OH proton, as can be seen in the barely resolved additional multiplicity in the lower trace. Coupling between CH—OH hydrogens is not always observed (Secs. 3.6 and 3.11), but it is here.

The mass spectrum in this example is unhelpful: there are no striking fragmentations and the only prominent peak, at m/z 55, is misleading, being a $C_4H_7^+$ fragment, which is not found intact in the parent structure. Aliphatic nitro compounds often decompose readily, either thermally or in the electron beam.

Example 9

The formula $C_8H_8O_2$ shows that five double-bond equivalents are present, and the strong absorption in the UV, with λ_{max} at 316 nm and an ε of 22 000, it is likely that four or five double bonds are conjugated to each other.

The IR spectrum shows that there are few, if any, saturated CHs (no strong bands just below 3000 cm^{-1}) and only some aryl or unsaturated CHs (the weak band at 3100 cm^{-1}). The carbonyl region shows two bands: one at 1695 cm^{-1} and the other at 1675 cm^{-1}. Since there is no nitrogen in the molecule, these cannot be from an amide, and so they are likely to be from an α,β-unsaturated ketone, aldehyde or acid. An acid is eliminated by the absence of the H-bonded OH absorption in the 3000-2500 cm^{-1} region, and an aldehyde is ruled out by the absence of absorption in the δ 9-10 region of the ^1H NMR spectrum. The compound is, therefore, probably a ketone. The strength and position of the band at 1615 cm^{-1} shows that a C=C conjugated double bond or conjugated aryl group must be present, and bands at 1555 (too weak to be a nitro group) and 1480 cm^{-1} also suggest an aromatic type of compound, although they are untypical enough to indicate that perhaps they are not a simple benzene ring.

The ^{13}C NMR shows eight signals—all, with the exception of a methyl group (the quartet at δ 27.83), in the unsaturated (δ >100) region, with a carbonyl carbon at δ 197.36. In the ^1H NMR spectrum, the sharp singlet at δ 2.31 suggests that the methyl group is part of a methyl ketone. This conclusion is supported by the presence in the mass spectrum of the base peak at M–15 (m/z 121). Moreover, the peaks m/z 94, 93 and 43 suggest the sequences **11** → **12** → **13**, **11** → **14** and **11** → **15**, all consistent with the presence of a methyl ketone.

$$[C_6H_5O]-COCH_3^{+\ddagger} \xrightarrow{\;-CH_3^{\cdot}\;} [C_6H_5O]-C\equiv O^+ \xrightarrow{\;-CO\;} [C_6H_5O]^+$$

11

$$\xrightarrow{\;-H_2C=C=O\;} [C_6H_6O]^{\ddagger}$$

$$\xrightarrow{\;-[C_6H_5O]^{\cdot}\;} COCH_3^+$$

12 m/z 121   **13** m/z 93

14 m/z 94

15 m/z 43

Spectra for example 9   C, 70.7%; H, 5.9%

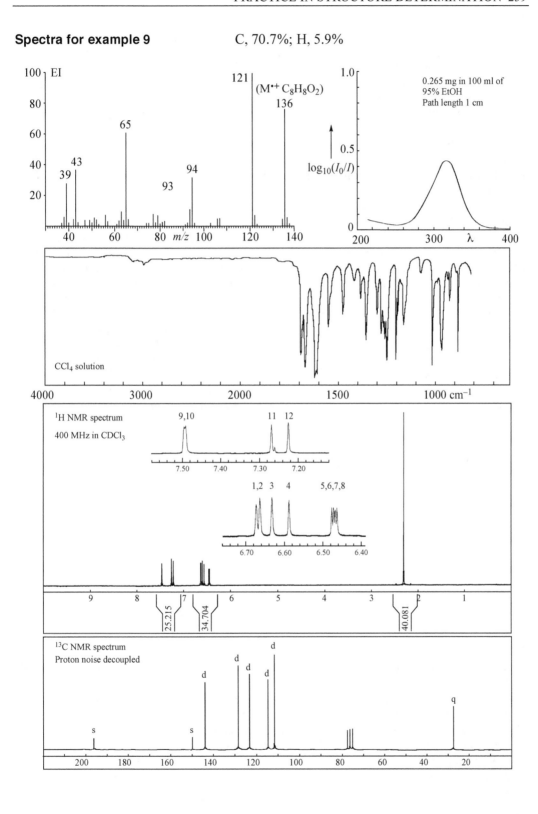

The ^{1}H NMR spectrum also shows that the remaining five hydrogen atoms are attached to double bonds, and the ^{13}C NMR shows that each of them is attached to a different carbon atom (because five of these signals are doublets, after allowing for the methyl quartet at δ 27.83). Eight lines can be seen in the lower superimposed trace, numbered 1-8, and a further four, numbered 9-12, in the upper superimposed trace. Lines 11 and 12 are a doublet at δ 7.28, with the relatively large coupling constant of 16 Hz, and they must be coupling to something in the eight-line pattern. Looking for a 16 Hz separation between the lines in this region reveals that only lines 3 and 4 have this separation. Since they are only a doublet, with no further coupling, there must be an AB system **17** present, and the coupling constant of 16 Hz shows that it must be a *trans*-disubstituted double bond. This, incidentally, is supported by the presence in the IR spectrum of a strong band at 970 cm^{-1}, a band typical of this feature (Table 2.2). We have now found fragments **16** and **17** (in which X and Y must carry no protons, since the AB system is not coupled to anything else). The remainder of the molecule is a C$_4$H$_3$O—unit, which does not have a carbonyl group (^{13}C NMR), does not have an OH (IR) and has each of the three hydrogens on a different carbon. With a little thought, the only possibility is seen to be a furan ring with one substituent **18**, and all we have left to deduce is whether that substituent is on C$_\alpha$ or C$_\beta$. The chemical shifts of the three remaining hydrogen atoms to be assigned (δ 7.50, 6.67 and 6.49) indicate that it is an α-substituted furan (one proton at low field and two at higher field; Table 3.24). The splittings of the three furan protons confirm this assignment: H$_\alpha$ is the fine doublet (lines 9 and 10) at δ 7.50 ($J_{\alpha\beta}$ = 1.5 Hz; Table 3.30), H$_{\beta'}$ is also a doublet (lines 1 and 2, $J_{\beta\beta'}$ = 3.5 Hz), and H$_\beta$ is a double doublet (lines 5, 6, 7, and 8, $J_{\beta\beta'}$ = 3.5 Hz, $J_{\alpha\beta}$ = 1.5 Hz).

The two carbonyl bands in the IR spectrum may be from the presence of the s-*trans* **19** and s-*cis* **20** conformers. Interconversion between **19** and **20** is rapid at room temperature, and the NMR spectra therefore present only time-averaged pictures.

16   17   18

19   20

Example 10

Mass spectra gave MH$^+$ (CI) 324 (C$_{15}$H$_{22}$N$_3$O$_5$ by high resolution) and fragment ions (EI) at 307 and 206. Amino acid analysis established the presence of threonine and tyrosine. The 2D COSY spectrum is presented here to the lower right of the diagonal, and a TOCSY spectrum to the upper left of the diagonal, with the diagonal taken from the COSY spectrum. The COSY spectrum has been recorded with a double quantum filter (DQF COSY), which removes (or greatly discriminates against) methyl and other singlets in the final spectrum. The normal 1D spectrum (singlet at δ 1.9 truncated to 75% of its true height) is given above the 2D spectrum. The spectra are recorded in d_3-methanol

(CD₃OH), containing *ca.* 5% H₂O to promote solubility of the sample, with almost complete removal of the signal from the RO*H* protons by presaturation of this resonance.

COSY and TOCSY spectra for example 10

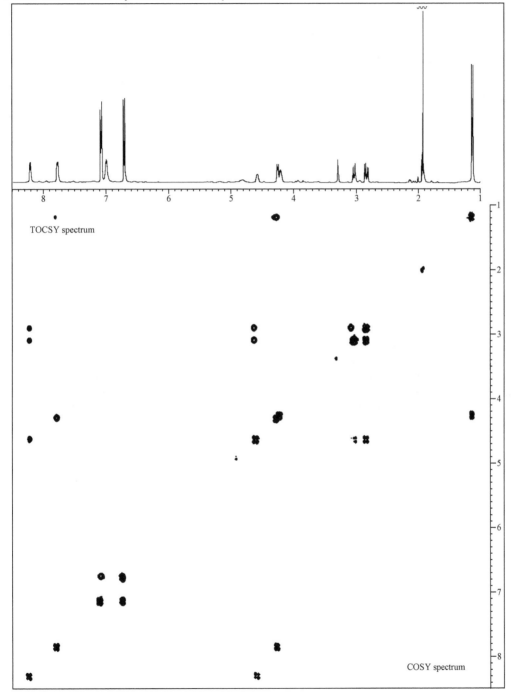

Since the in-chain masses of threonine and tyrosine are 101 and 163 (Table 4.13), the mass unaccounted for is 59 Daltons (C_2H_5NO). The threonine and tyrosine spin systems can be identified in the proton NMR spectra. For tyrosine **21**, the CH_2 protons (non-equivalent, because they are adjacent to a stereogenic centre) at δ 3.04 and 2.82, both coupled to the tyrosine α-CH at δ 4.58, which is further coupled to the tyrosine NH at δ 8.21. The *ortho*-coupled (8 Hz) protons of the tyrosine aromatic ring are at δ 6.70 and 7.09. For threonine **22**, its β-methyl resonance is at δ 1.13, coupled to the CH(OH) proton at δ 4.20. This latter proton is barely resolved from the threonine α-CH at δ 4.25. The coupling between the δ 4.25 and 4.20 signals is too near the diagonal to be observed in the 2D spectra. The α-CH at δ 4.25 is coupled to the threonine NH at δ 7.78. Note how the TOCSY spectrum directly picks up the fact that the tyrosine CH_2-protons correlate with the NH at δ 8.21 and that the threonine CH_3-protons correlate with the NH at δ 7.78.

21   **22**

Unassigned signals that correspond to one or more protons are at δ 6.98 (*ca.* 2H), 4.80 (lack of 100% removal of ROH protons in the presaturation), 3.3 (trace of CHD$_2$OH in CD$_3$OH) and 1.91 (3H). The last of these corresponds closely to the CH_3 of an *N*-acetyl group (CH_3CON at δ 2.0 ±0.2, Table 3.19). Since fragment ions occur at 206 and 307 in the mass spectrum, these are in accord with N-terminal acetylation of a dipeptide and formation of the acylium ions identified on the drawing **23**, thus indicating the sequence *N*-acetyl-Tyr-Thr-. The missing mass is therefore an NH_2 group, which gives rise to the signal at δ 6.98.

23

Example 11

Example 11 illustrates the power of HMBC spectra (long-range ^{1}H-^{13}C COSY) (Sec. 3.21), especially where information about proton-proton connectivity is sparse, typically because of a high degree of substitution or unsaturation, or both.

Spectra for example 11

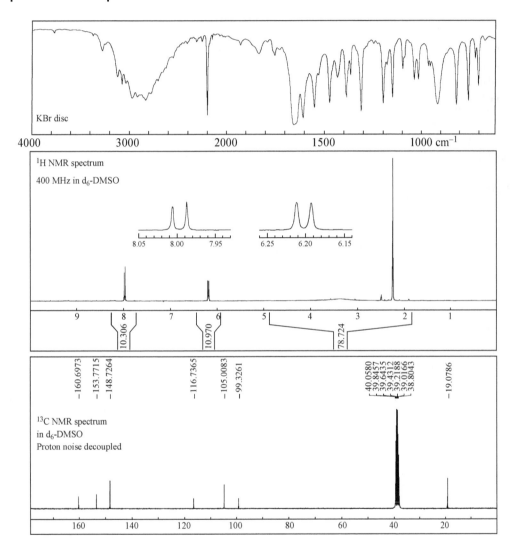

The molecular formula from HRMS is $C_7H_6N_2O$. There are therefore six DBEs. The IR spectrum shows clear evidence for a nitrile (2225 cm^{-1}), and at first glance might suggest a carboxylic acid (the broad band in the region 2500-3300 cm^{-1}). However, the latter possibility is precluded by the molecular formula. An intense band at 1660 cm^{-1} is consistent with the presence of an amide (Table 2.7). The ^{13}C spectrum is taken in d$_6$-DMSO solution with the solvent giving rise to the multiplet at δ 39. It shows seven chemically distinct carbon atoms, and judging from the peak intensities, three of these (δ

149, 105 and 19) might tentatively be inferred to carry directly bonded protons while the others do not.

The proton spectrum is also taken in d₆-DMSO solution with the solvent giving rise to the singlet at δ 2.50. It shows two vicinally coupled protons in the region for trigonally bound protons, and a methyl group at δ 2.3, which indicates that it too is probably attached to a trigonal carbon. Since the vicinal coupling constant is *ca.* 8 Hz, a sensible working hypothesis is that they are *cis*-oriented as part of a six-membered ring. On the basis of the molecular formula, there remains only one proton to be identified, but the very broad signal centred at δ 3.4 integrates for almost four protons. It seems likely that most of this signal is from traces of water in the DMSO, which may arise either because the solvent contained traces of water prior to addition of sample or because the sample itself contained traces of water. The chemical shift of the broad signal is consistent with this interpretation, since traces of water in DMSO resonate at δ 3.3 (Table 3.26). Thus, the remaining proton of the sample is one that can exchange with water, and is doing so at such a rate that it is broadening the water signal. This may also explain the broad strong band in the IR spectrum, and given that we think that there may be an amide group it seems likely that this proton is that of the N*H* of an amide.

HMBC spectrum for example 11

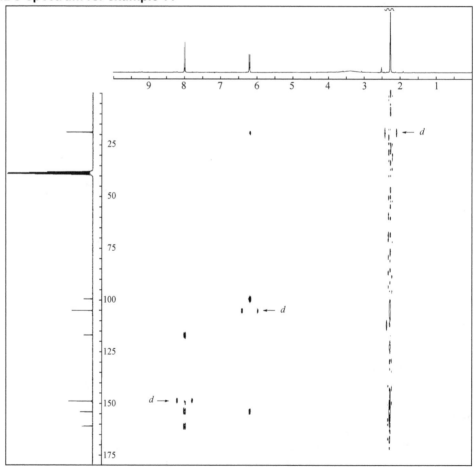

The HMBC spectrum serves not only its intended function of indicating protons and carbons separated by two or three bonds, but also correlates protons to the carbons to which they are directly attached, serving the function of an HMQC spectrum (Sec. 3.21). These direct attachments are exposed by cross-peaks which occur as doublets, labelled *d* in the HMBC spectrum), with the doublet splitting occurring in the proton dimension and corresponding to the direct ^{13}C—^{1}H coupling constant. Thus, doublets indicate the direct attachment of protons at δ 8.0, 6.2 and 2.3 to carbons at δ 149, 105 and 19, respectively. The singlet cross-peaks indicate two- or three-bond correlations. Note that no such correlations can be inferred for the CH$_3$ protons, because the spectrum has too much noise at the chemical shift of these protons. The occurrence of such noise is not uncommon at the chemical shift of intense methyl group signals.

The structural entities so far inferred are therefore those indicated in the box above. A singlet cross-peak in the HMBC spectrum shows that the carbon atom of the methyl group δ 19 must be separated by two or three bonds from the proton at δ 6.2. The methyl group must therefore be either W or X, and only the former if the assumption of the presence of a six-membered ring is correct. The nitrile carbon must be the signal at δ 117, within the usual range δ 110–125 (Table 3.9). A singlet cross-peak shows that it must be separated by two or three bonds from the proton at δ 8.0, and the nitrile therefore corresponds to Z. From the molecular formula, X—Y corresponds to a secondary amide (NH—CO). The chemical shifts can only be satisfied by X = NH and Y = CO, with an enamine-type polarisation to account for the relatively high-field chemical shifts of the carbons at δ 105 and the proton at δ 6.2. These signals would have been expected at lower field if the CO and NH groups had been the other way round. All the available spectroscopic evidence is therefore satisfied by the α-pyridone structure **24**.

24

5.5 Simple problems using ^{13}C NMR or joint application of IR and ^{13}C NMR

Problem 1

Three compounds of molecular formula C_4H_8O gave the following ^{13}C and IR data.

Compound **25**:	IR: 1730 cm^{-1}
	^{13}CNMR: δ 201.6, 45.7, 15.7 and 13.3
Compound **26**:	IR: 3200 (broad) cm^{-1}
	^{13}CNMR: δ 134.7, 117.2, 61.3 and 36.9
Compound **27**:	IR: no peaks except CH and fingerprint
	^{13}C NMR: δ 67.9 and 25.8

Suggest a structure for each compound, and then see whether your suggestions are compatible with the following information. Compound **25** reacts with NaBH$_4$ to give compound **28**, $C_4H_{10}O$, IR 3200 (broad) cm^{-1} and ^{13}C NMR δ 62.9, 36.0, 20.3 and 15.2. Compound **26** reacts with hydrogen over a palladium catalyst to give the same product **28**, while compound **27** reacts with neither reagent.

Problem 2

Given the spectroscopic data, deduce structures for the compounds **29**, **30** and **31**.

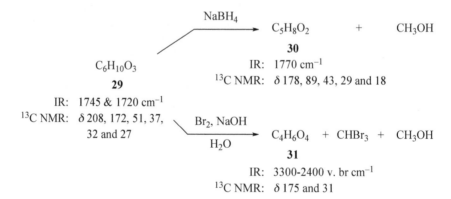

Problem 3

The two-dimensional ^{13}C spectrum shown below, with the ^{13}C spectrum on both axes, is of the methyl region of the heptamethylbenzene cation **32** in concentrated sulfuric acid solution under conditions where methyl group migration is occurring rapidly. Is the methyl group migration 1,2 or 1,3 or 1,4 or random? In a multi-pulse experiment, if a nucleus is excited in chemical environment a, and during the course of the experiment it moves into chemical environment b (through a rearrangement), then a cross-peak can be observed between the chemical shifts associated with the two environments. The 2D spectrum showing this effect is described as an 'exchange spectrum'.

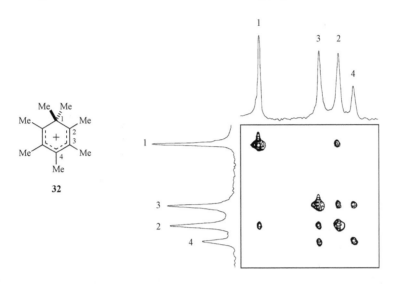

32

5.6 Simple problems using ^{1}H NMR

Problem 4

Two isomeric compounds C_4H_8O gave the following ^{1}H NMR data, in which J values <1.5 Hz were not resolved.

Compound **33**: δ 3.8 (2H, s), 3.5 (3H, s) and 1.8 (3H, s)

Compound **34**: δ 4.95 (1H, broad s), 4.8 (1H, broad s), 4.0 (2H, s), 2.2 (1H, s, removed on shaking with D_2O) and 1.7 (3H, s)

Deduce their structures.

Problem 5

Iodolactonisation of the acid **35** gives a single product **36**, the ^{1}H NMR spectrum of which contains the signals listed. What is the stereochemistry of compound **36**?

δ 3.64 (1H, ddd, J = 9.7, 3.4 and 2.9 Hz)
δ 3.55 (1H, dd, J = 11.2 and 2.9 Hz)
δ 3.41 (1H, dd, J = 11.2 and 3.4 Hz

35   **36**

Problem 6

Compounds **37** and **38** are produced by the following reaction scheme. Deduce the structure of the final product **38**, including the relative stereochemistry and the conformation.

δ 3.5 (1H, d, J = 12 Hz)
δ 3.3 (1H, dd, J = 12 and 2 Hz)
δ 2.6 (1H, dqd, J = 12, 7 and 6 Hz)
δ 2.1 (1H, ddd, J = 13, 6 and 2 Hz)
δ 1.8 (1H, dd, J = 13 and 12 Hz)
δ 1.4 (3H, s)
δ 1.3 (3H, d, J = 7 Hz)

The ^{1}H NMR spectrum of compound **38** also showed one exchangeable proton and five aromatic protons. Irradiation of either the signal at δ 3.3 or at 2.1 gives a small positive enhancement of the signal at δ 1.4.

5.7 Problems using a combination of spectroscopic methods

Problem 7

Three compounds, **39-41**, of formula C_4H_6 have been prepared by the routes shown below. Suggest structures for these isomers.

39   ^{1}H NMR: δ 5.35 (2H, s) and 1.0 (4H, s)
^{13}C NMR: 3 different signals

40   ^{1}H NMR: δ 6.4 (1H, t, J = 1 Hz), 2.13 (3H, s)
and 0.84 (2H, d, J = 1 Hz)
^{13}C NMR: 4 different signals

41   ^{1}H NMR: δ 7.18 (2H, d, J = 1 Hz), 1.46 (1H, qt, J = 5 and 1 Hz)
and 0.97 (3H, d, J = 5 Hz)
^{13}C NMR: 3 different signals

Compound **39** reacts with *m*-chloroperbenzoic acid to give compound **42**, which rearranges to compound **43** in the presence of lithium iodide. Suggest structures for compounds **42** and **43**.

39 $\xrightarrow{\textit{m}\text{-ClC}_6\text{H}_4\text{CO}_3\text{H}}$ **42** (C_4H_6O) IR: no strong peaks except in the fingerprint region
^{1}H NMR: δ 3.0 (2H, s) and 0.90-0.80 (4H, A_2B_2 system)

42 $\xrightarrow{\text{LiI}}$ **43** (C_4H_6O) IR: ν_{max} (cm^{-1}) 1770
^{1}H NMR: δ 3.02 (4H, t, $J=5$ Hz) and 1.98 (2H, q, $J=5$ Hz)

Problem 8

Deduce structures for the products **44-46** of the following reactions.

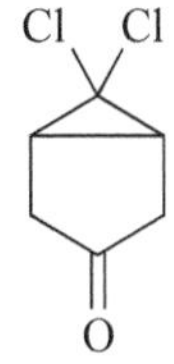

$\xrightarrow[\text{44}]{\text{H}_2\text{O, 100 °C}}$ MS: M$^{\bullet+}$ 128 and 130
IR: ν_{max} (cm^{-1}) 3500, 1600 and 1500
^{1}H NMR: δ 7.1 (2H, d, $J=7$ Hz), 6.8 (2H, d, $J=7$ Hz) and 5.4 (1H, br s)

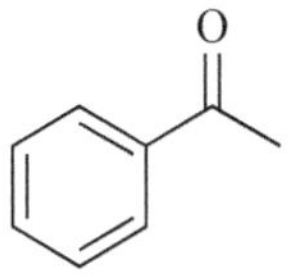

$\xrightarrow[\text{45}]{\text{NaOH}}$ MS: M$^{\bullet+}$ 306
IR: ν_{max} (cm^{-1}) 1600 and 1500
^{1}H NMR: δ 8.05 (3H, s), 7.64 (6H, d, $J=6$ Hz) and 7.5-7.3 (9H, m)
^{13}C NMR: δ 142, 141, 129, 127, 126 and 125 (142 and 141 are weak)

$\xrightarrow[\text{46}]{\text{AlCl}_3}$ MS: M$^{\bullet+}$ 110
IR: ν_{max} (cm^{-1}) 1720
^{1}H NMR: δ 7.7 (1H, dt, $J=6$ and 3 Hz), 6.2 (6H, dt, $J=6$ and 2 Hz),
2.2 (1H, dd, $J=3$ and 2 Hz) and 1.1 (6H, s)

Problem 9

The bis(bromomethyl)benzil **47** reacts with sodium hydroxide to give a product **48**. Deduce its structure.

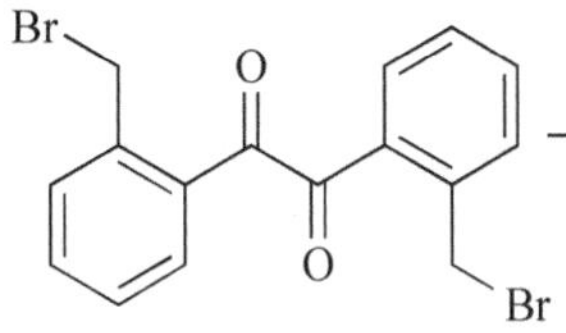

$\xrightarrow{\text{NaOH}}$ **48** $C_{16}H_{12}O_3$ IR: ν_{max} (cm^{-1}) 1700
^{1}H NMR: δ 8.0-7.0 (8H, m), 5.5 (1H, d, $J=16$ Hz),
5.3 (1H, d, $J=13$ Hz), 5.2 (1H, d, $J=13$ Hz) and 4.9 (1H, d, $J=16$ Hz)
^{13}C NMR: δ 189(s), 12 aromatic carbons (8d and 4s), 109(s), 74(t) and 63(t)

47

Problem 10

Given the following spectroscopic and chemical data, deduce a structure for a toxic substance **49**, $C_{11}H_{16}N_2O_5$, isolated from the Colorado potato beetle.

MS (FIB): m/z 257; UV: λ_{max} (nm) 230 (ε 8000); IR: ν_{max} (cm^{-1}) 3400-2500 (br), 1660 and 1590.

^{13}C NMR

δ	Multiplicity	δ	Multiplicity	δ	Multiplicity
178.6	s	133.2	d	54.9	d
175.6	s	127.7	d	33.4	t
175.2	s	122.4	t	28.0	t
135.1	d	56.5	d		

^{1}H NMR (in D$_2$O)

δ	Intensity	Multiplicity	J (Hz)
6.75	1H	dt	16.7 and 10.2
6.26	1H	t	10.2
5.43	1H	dd	10.2 and 9.6
5.38	1H	dd	16.7 and 1.8
5.29	1H	dd	10.2 and 1.8
5.06	1H	d	9.6
3.74	1H	t	6.0
2.45	2H	t	7.5
2.13	2H	m	

Hydrogenation of the toxin with H$_2$ on Pd/C gave a compound whose ammonia CI spectrum showed m/z 261 and fragment ions at 132 and 130. When CF$_3$CO$_2$H was added to a D$_2$O solution of this reduced product, a triplet in its proton NMR spectrum shifted from δ 3.8 (1H, J = 6 Hz) to δ 4.2. This same reduced product, on treatment with 2,4-dinitrofluorobenzene, followed by acid hydrolysis (6N HCl, 373 K, 12 h) gave one molar equivalent of the amino acid **50**.

50

In the following 23 problems, which start with simple examples and go on to moderately difficult ones, the actual spectra are reproduced, and the conditions under which the spectra were obtained are indicated on the spectra. Unless otherwise stated, no changes were observed in the NMR spectra after the solutions had been shaken with D$_2$O. For those needing further practice, and access to a host of real problems, the *Journal of Antibiotics* is a rich source (http://www.antibiotics.or.jp/). The evolution of the methodology of structure elucidation—from the inception of the journal to the present time—chronicles the ever-changing face of the subject. The most recent volumes illustrate the combination of methods that is currently the most appropriate. The complexity of the structures attests to the power of current methods, and their diversity continues to challenge the mind of the synthetic chemist.

Problem 11

C, 80.7%; H, 7.6%

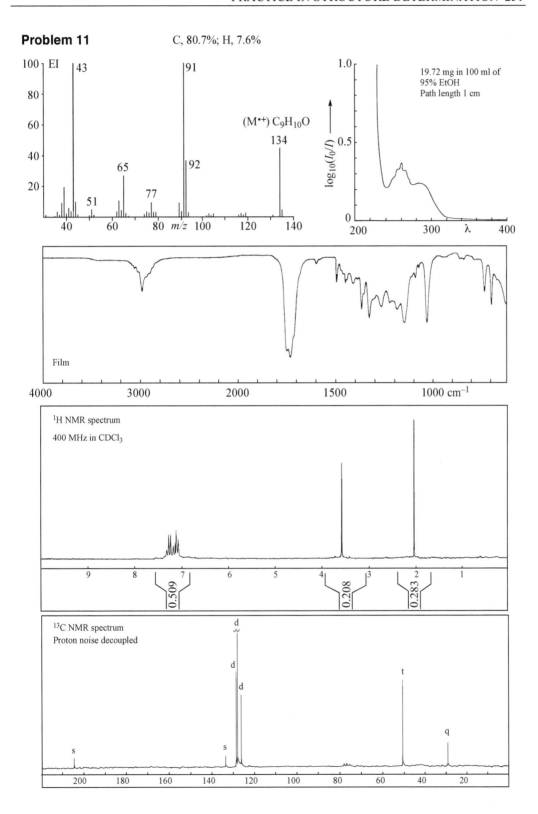

$(M^{\bullet+})\ C_9H_{10}O$

134

19.72 mg in 100 ml of
95% EtOH
Path length 1 cm

Film

^{1}H NMR spectrum

400 MHz in CDCl$_3$

0.509   0.208   0.283

^{13}C NMR spectrum
Proton noise decoupled

Problem 12   C, 49.4%; H, 9.8%; N, 19.1%

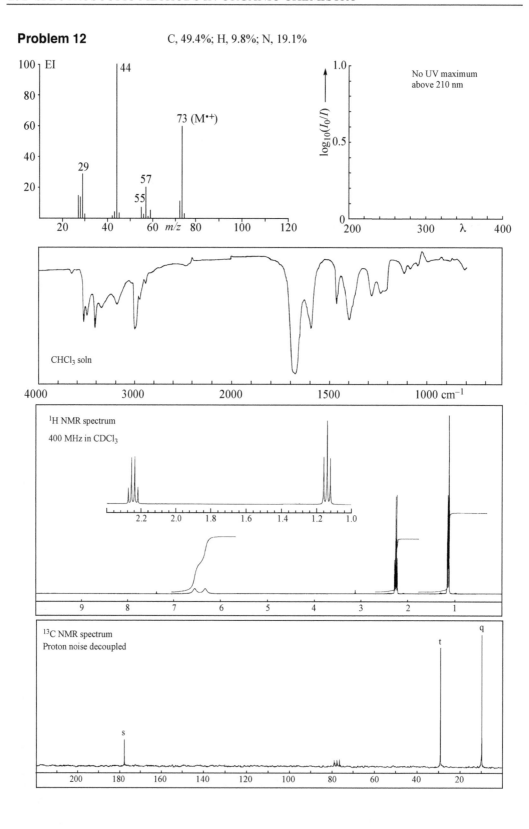

Problem 13

C, 80.0%; H, 4.8%

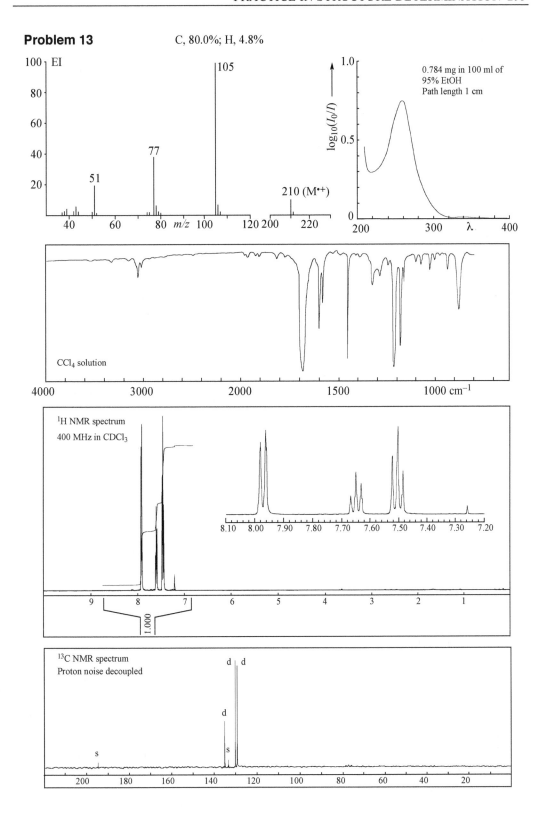

Problem 14

C, 64.7%; H, 10.9%

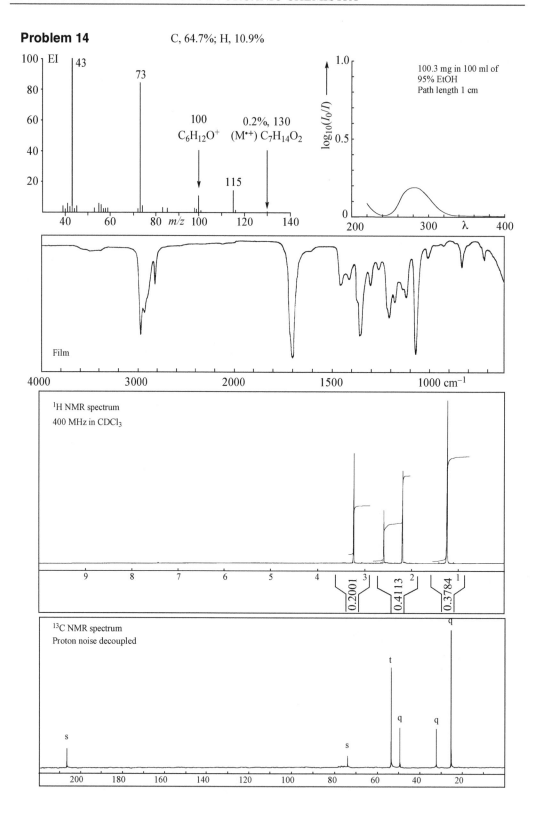

100.3 mg in 100 ml of
95% EtOH
Path length 1 cm

100
$C_6H_{12}O^+$

0.2%, 130
(M⁺•) $C_7H_{14}O_2$

115

Film

^{1}H NMR spectrum
400 MHz in CDCl$_3$

0.2001   0.4113   0.3784

^{13}C NMR spectrum
Proton noise decoupled

Problem 15

C, 58.2%; H, 8.5%

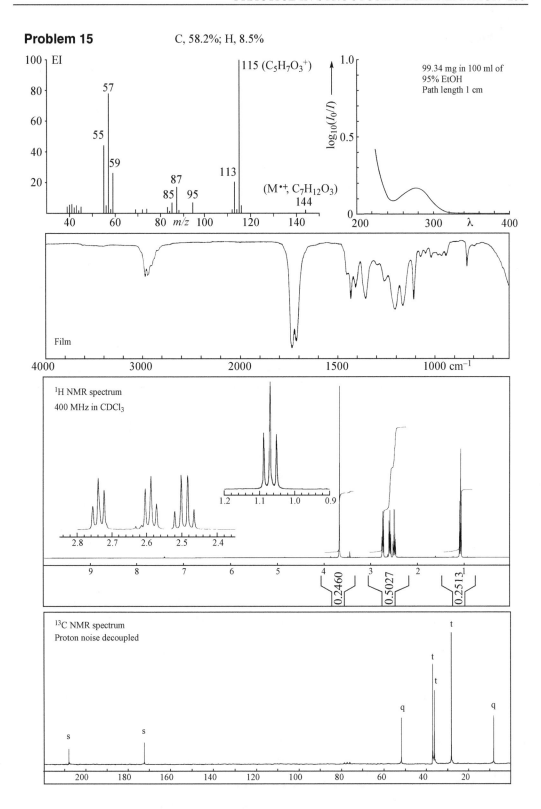

Problem 16

C, 36.5%; H, 10.0%

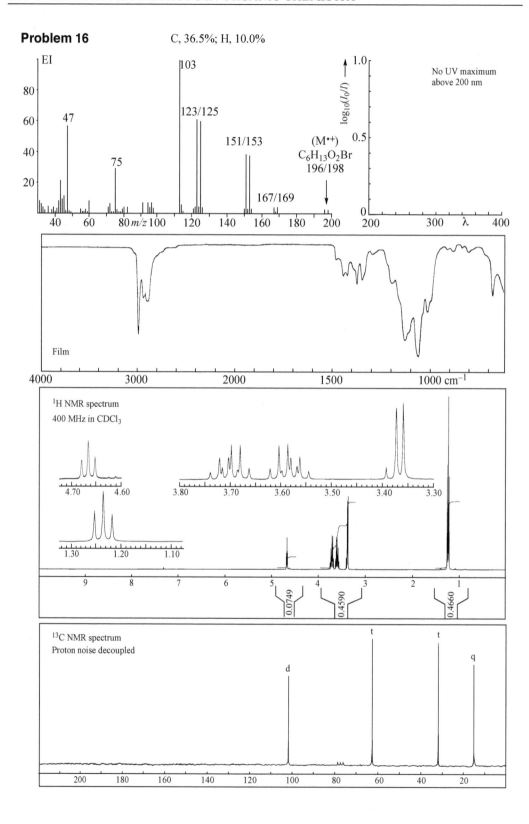

Problem 17   C, 39.8%; H, 7.3%

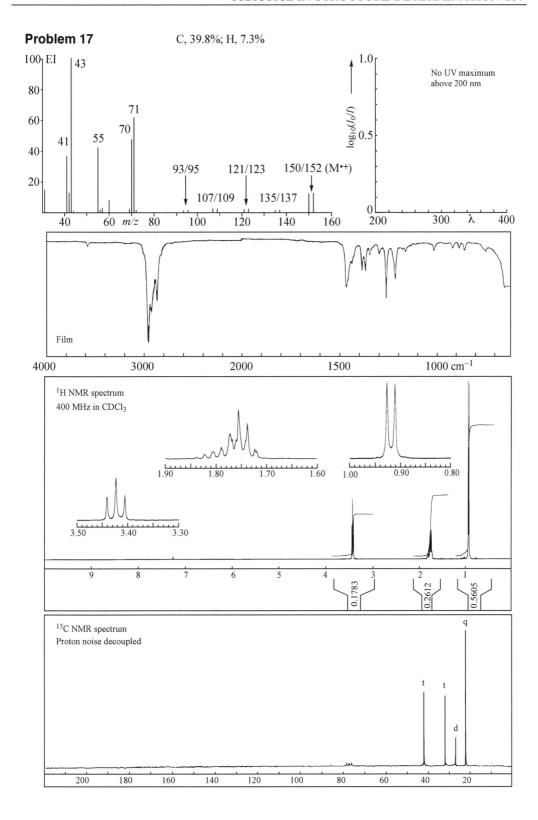

Problem 18

C, 64.3%; H, 8.8%

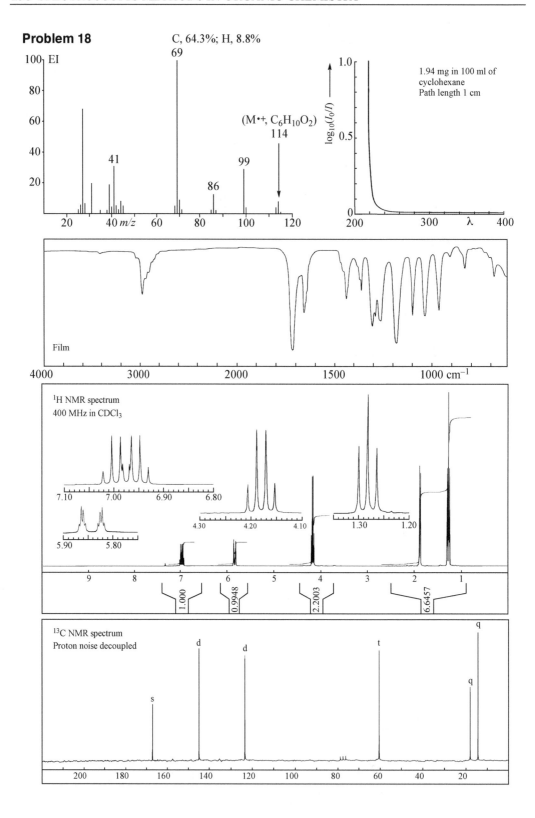

Problem 19

C, 55.1%; H, 4.6; N, 9.1%

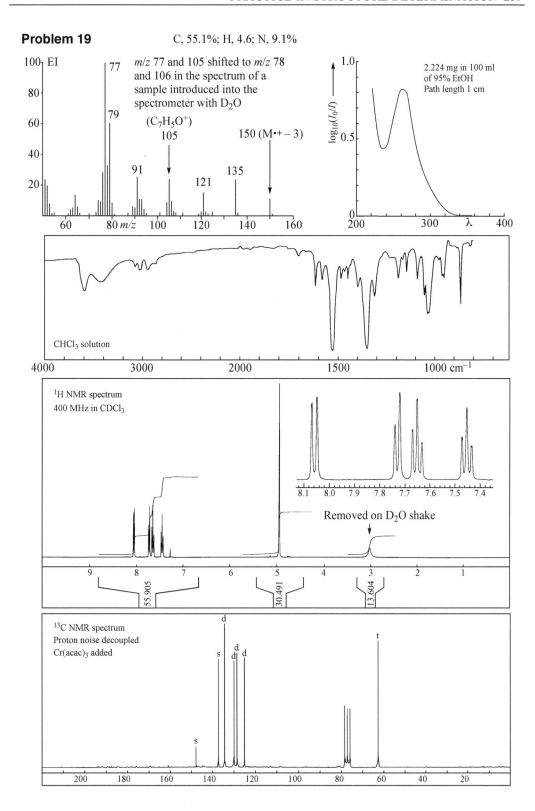

m/z 77 and 105 shifted to *m/z* 78 and 106 in the spectrum of a sample introduced into the spectrometer with D_2O

$(C_7H_5O^+)$

150 $(M^{\bullet+} - 3)$

2.224 mg in 100 ml of 95% EtOH
Path length 1 cm

CHCl₃ solution

¹H NMR spectrum
400 MHz in CDCl₃

Removed on D₂O shake

¹³C NMR spectrum
Proton noise decoupled
Cr(acac)₃ added

Problem 20

C, 67.7%; H, 6.4; N, 8.0%

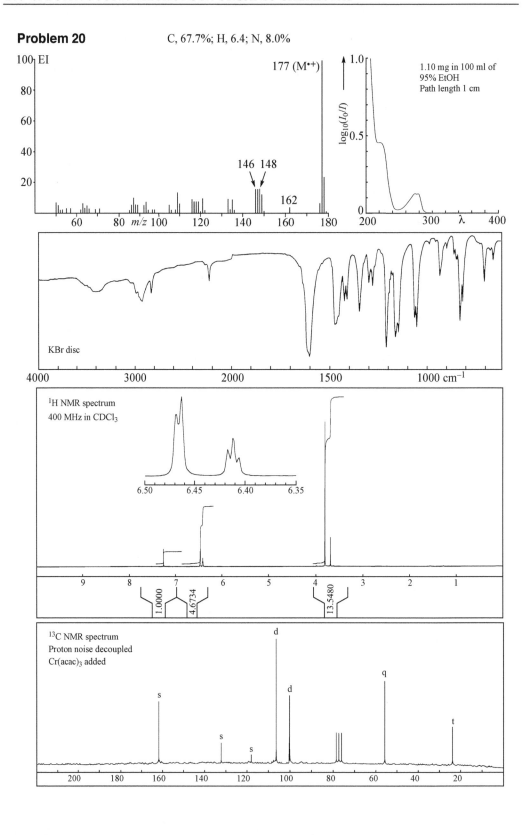

100 EI

177 (M•+)

146 148

162

1.0

$\log_{10}(I_0/I)$

0.5

1.10 mg in 100 ml of
95% EtOH
Path length 1 cm

200   300   λ   400

60   80   m/z 100   120   140   160   180

KBr disc

4000   3000   2000   1500   1000 cm⁻¹

¹H NMR spectrum
400 MHz in CDCl₃

6.50   6.45   6.40   6.35

9   8   7   6   5   4   3   2   1

1.0000   4.6734   13.5480

¹³C NMR spectrum
Proton noise decoupled
Cr(acac)₃ added

s   s   s   d   d   q   t

200   180   160   140   120   100   80   60   40   20

Problem 21   C, 45.2%; H, 4.2; N, 7.5%

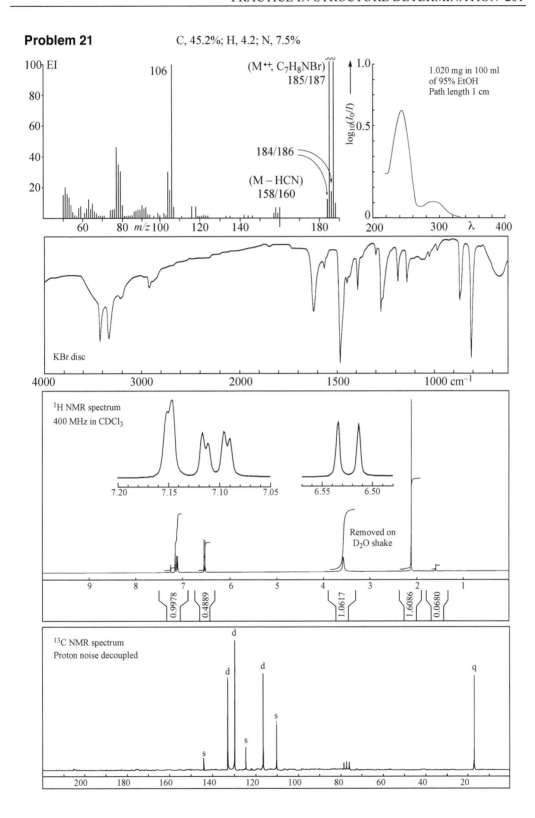

Problem 22

C, 75.5%; H, 7.5; N, 8.1%

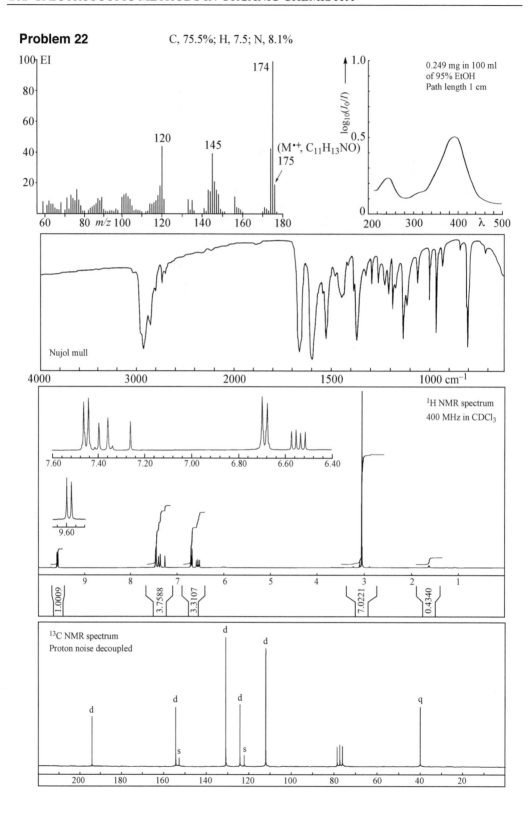

Problem 23   C, 63.7%; H, 6.5; N, 29.9%

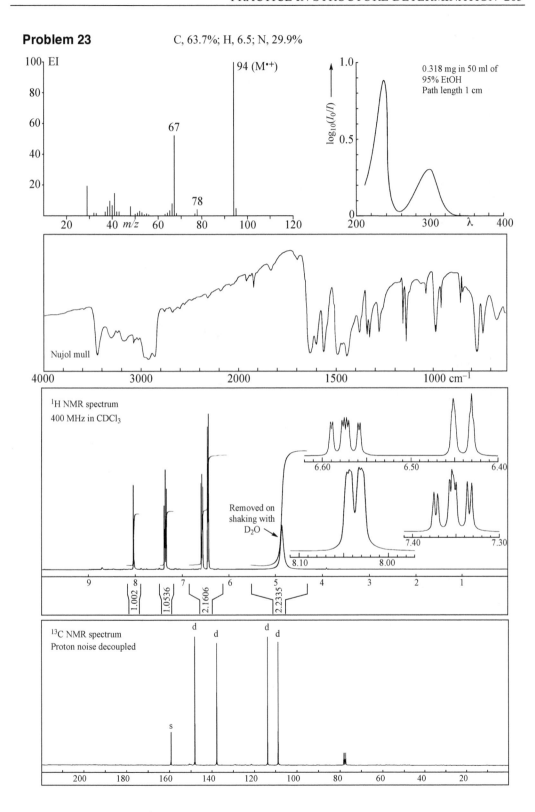

^{1}H NMR spectrum
400 MHz in CDCl$_3$

Removed on
shaking with
D$_2$O

^{13}C NMR spectrum
Proton noise decoupled

0.318 mg in 50 ml of
95% EtOH
Path length 1 cm

Nujol mull

Problem 24   C, 51.1%; H, 2.8; N, 10.0%

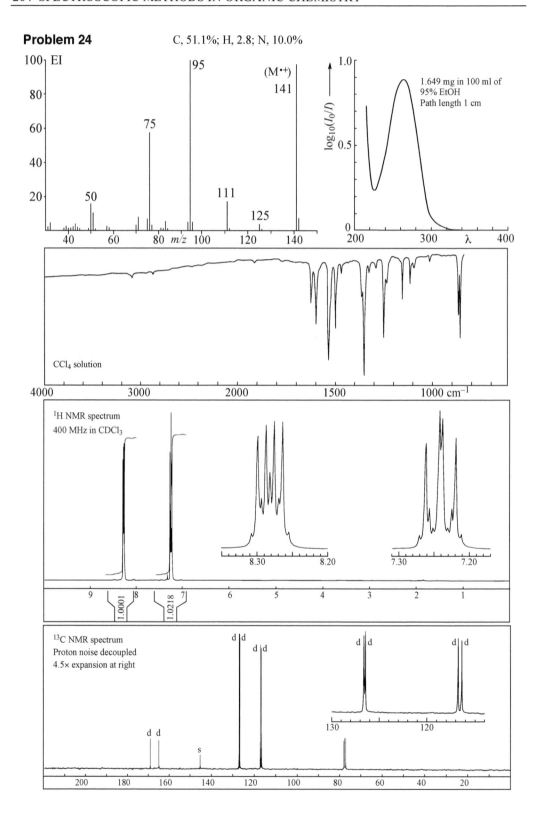

Problem 25

C, 41.7%; H, 2.0; N, 7.0%

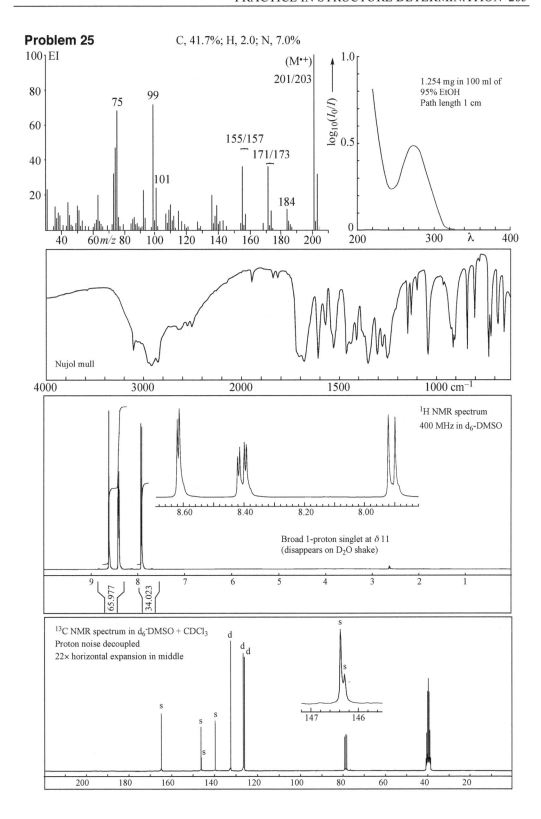

(M•+) 201/203

155/157

171/173

184

99

75

101

1.254 mg in 100 ml of 95% EtOH
Path length 1 cm

Nujol mull

¹H NMR spectrum
400 MHz in d₆-DMSO

Broad 1-proton singlet at δ 11
(disappears on D₂O shake)

65.977

34.023

¹³C NMR spectrum in d₆-DMSO + CDCl₃
Proton noise decoupled
22× horizontal expansion in middle

Problem 26

C, 68.7%; H, 6.3%

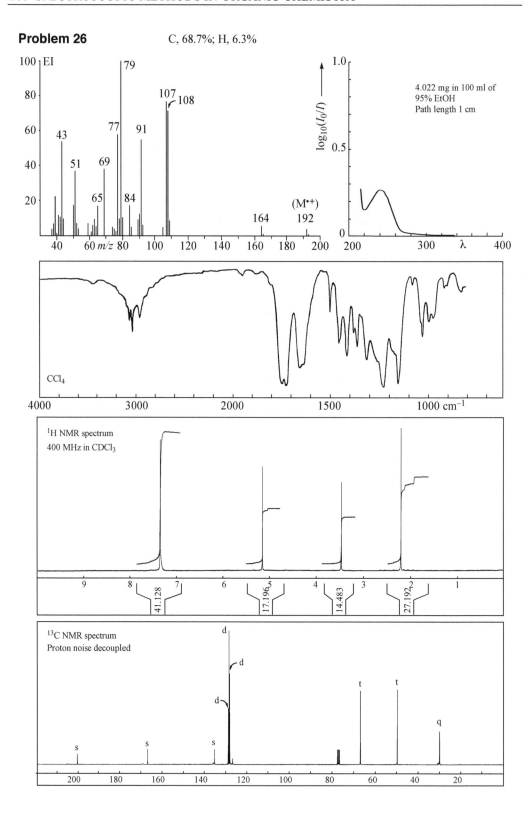

Problem 27   C, 67.2%; H, 4.6; N, 13.1%

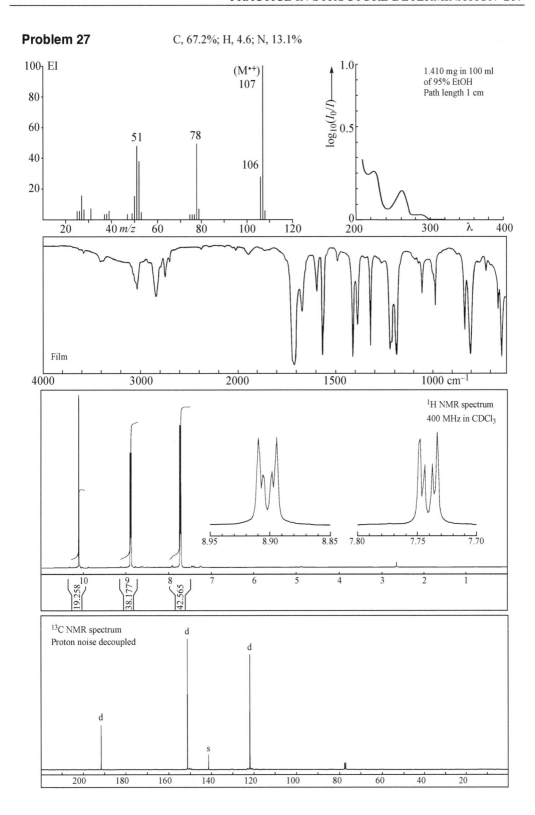

Problem 28

C, 68.1%; H, 7.2%

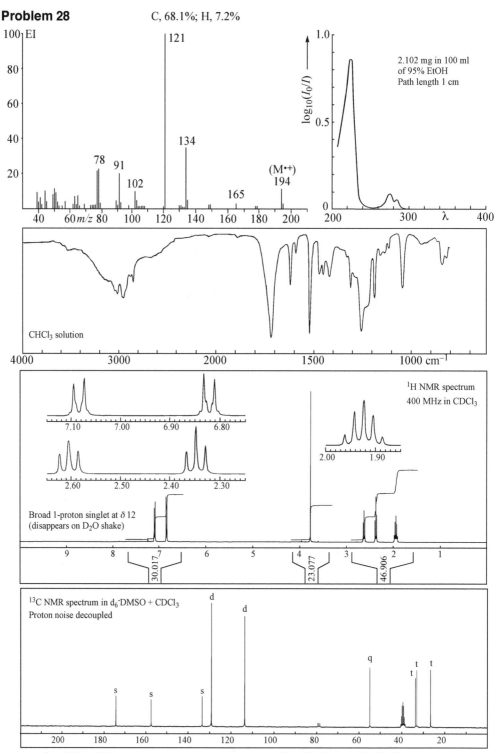

Problem 28 continues on the next page with HMQC and COSY spectra

Problem 28 (continued)

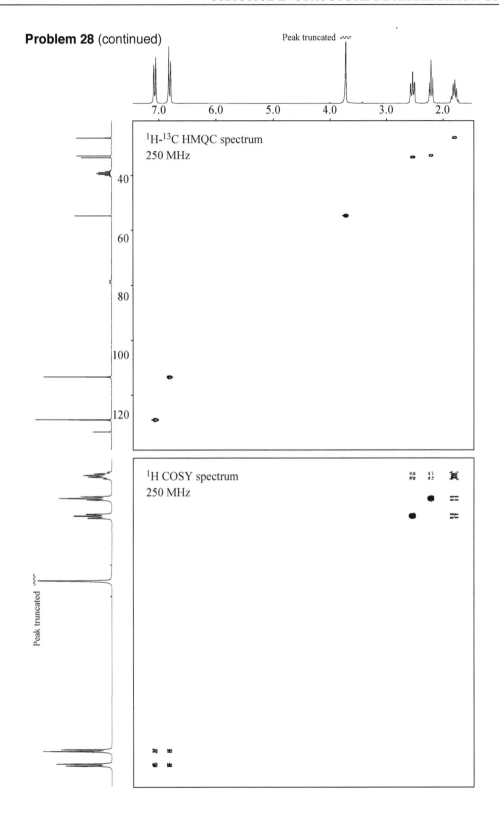

Peak truncated ∿

^{1}H-^{13}C HMQC spectrum
250 MHz

^{1}H COSY spectrum
250 MHz

Peak truncated

Problem 29

C, 54.8%; H, 4.8; N, 9.3%

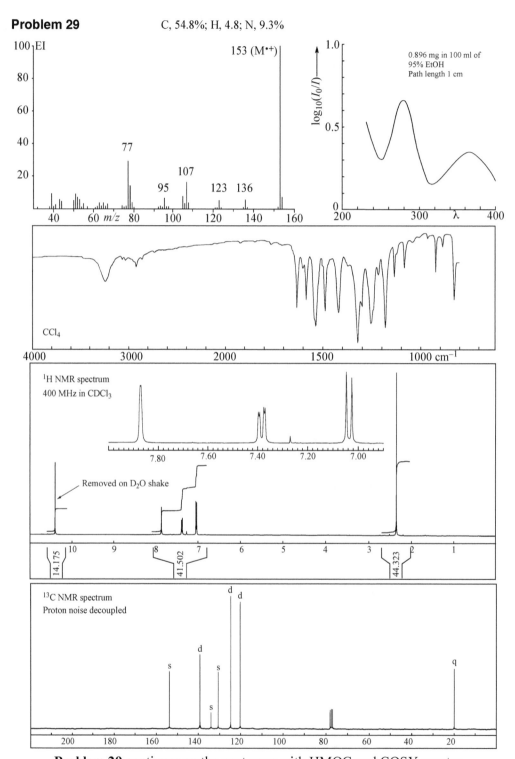

Problem 29 continues on the next page with HMQC and COSY spectra

Problem 29 (continued)

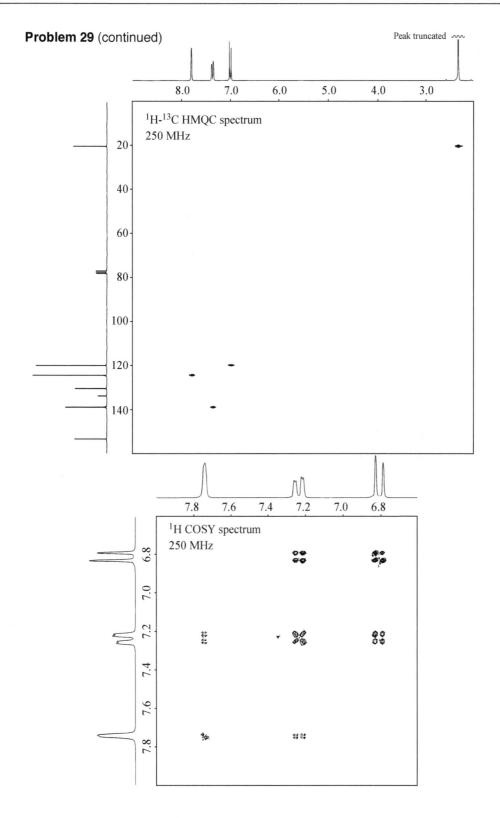

Peak truncated

^{1}H-^{13}C HMQC spectrum
250 MHz

^{1}H COSY spectrum
250 MHz

Problem 30   C, 71.8%; H, 6.8%

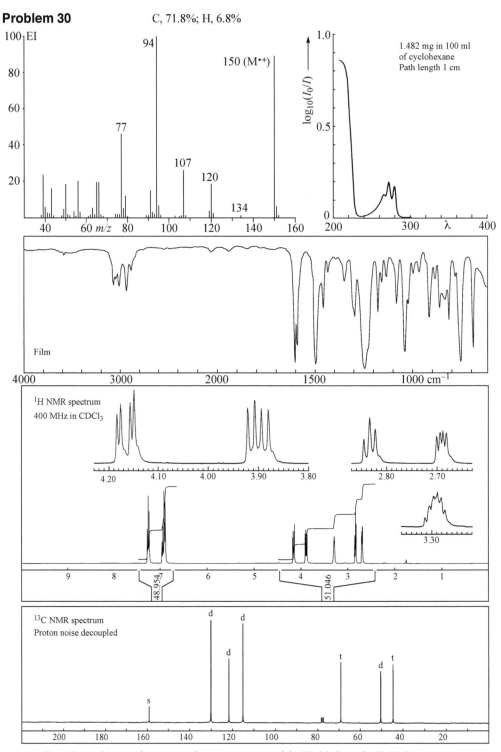

Problem 30 continues on the next page with HMQC and HMBC spectra

Problem 30 (continued)

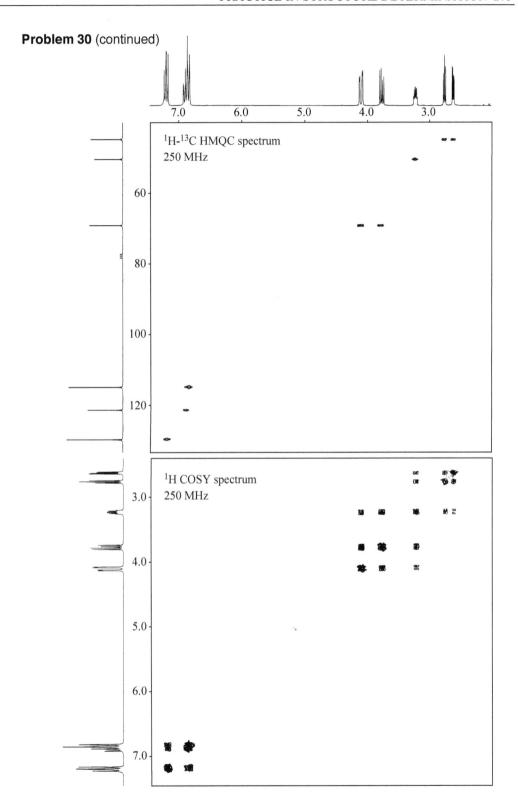

^{1}H-^{13}C HMQC spectrum
250 MHz

^{1}H COSY spectrum
250 MHz

Problem 31

High-resolution MS establishes the molecular formula as $C_6H_{11}NO$. There is no UV absorption maximum above 200 nm. There are intense IR bands at 1680 and 1665 cm^{-1} (solid state).

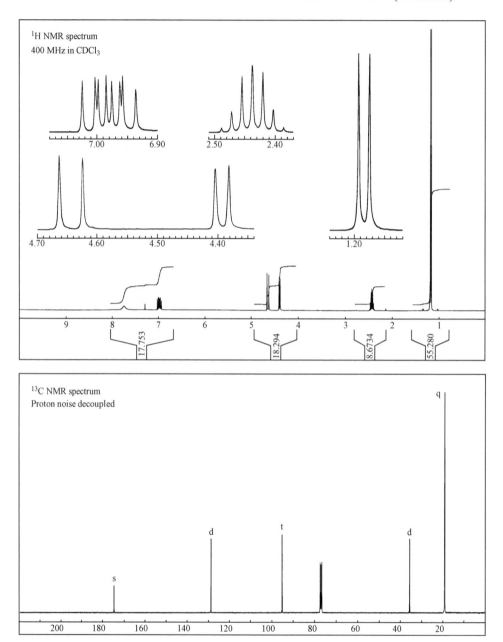

Problem 31 continues on the next page with HMQC and HMBC spectra

Problem 31 (continued)

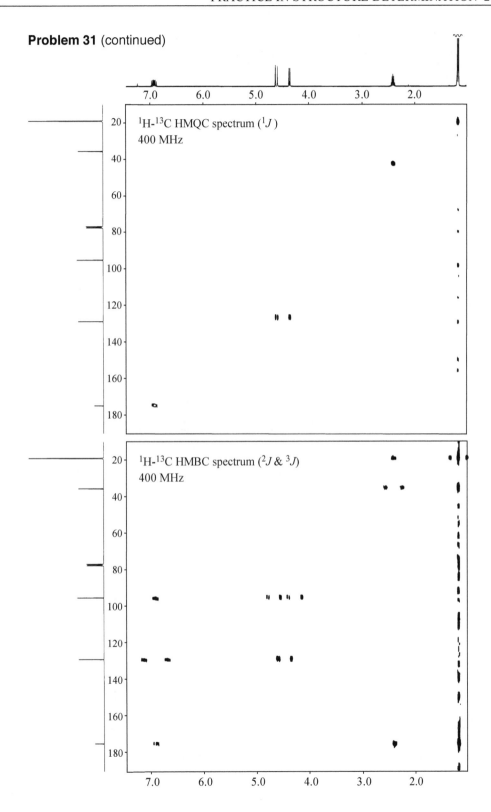

^{1}H-^{13}C HMQC spectrum (^1J)
400 MHz

^{1}H-^{13}C HMBC spectrum (^2J & ^3J)
400 MHz

Problem 32

High-resolution MS establishes the molecular formula as $C_{10}H_{15}NO_5$. There is no UV absorption maximum above 200 nm. There are intense IR bands at 3330, 1740, 1715, 1670, 1638 and 1545 cm^{-1} (solid state).

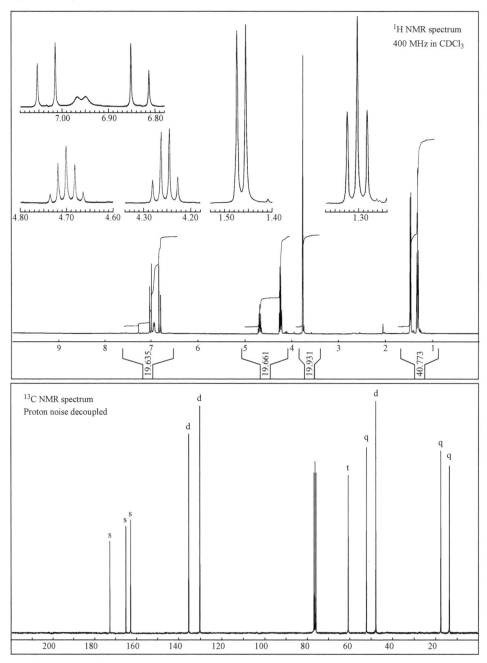

Problem 32 continues on the next page with HMQC and HMBC spectra

Problem 32 (continued)

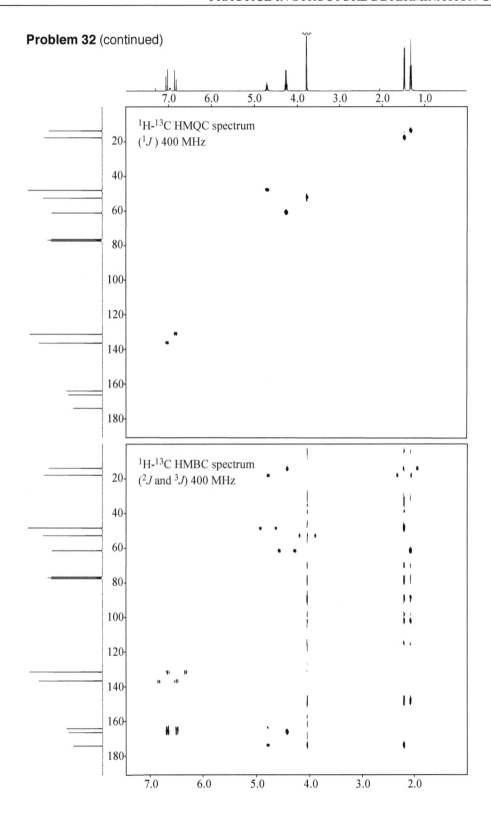

Problem 33   C, 69.6%; H, 11.6%

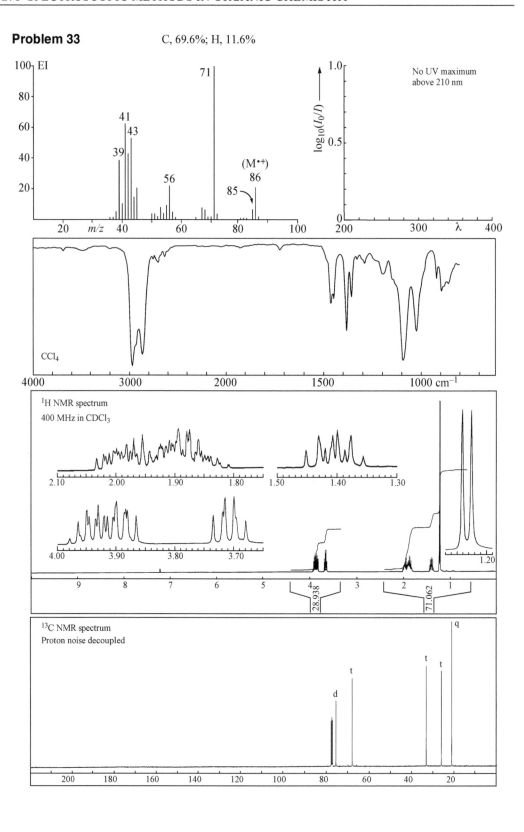

No UV maximum above 210 nm

5.8 Answers to problems 1-33

Problems 1 and 2: **25** butanal; **26** but-3-en-1-ol; **27** tetrahydrofuran; **28** butan-1-ol; **29** methyl 4-oxopentanoate; **30** 5-methyldihydrofuran-2(3H)-one; **31** butanedioic acid (succinic acid).

Problem 3: 1,2 Migration (although this topic was not discussed in the NMR chapter, it should be obvious enough how to use cross-peaks from the way the cross-peaks from the other 2D techniques are interpreted)

Problems 4, 5 and 6: **33** 2-methoxyprop-1-ene; **34** 2-methylprop-2-en-1-ol;

36

38

Problems 7, 8, 9 and 10: **39** methylenecyclopropane; **40** 1-methylcycloprop-1-ene; **41** 3-methylcycloprop-1-ene; **42** 1-oxaspiro[2.2]pentane; **43** cyclobutanone; **44** 4-chlorophenol; **45** 1,3,5-triphenylbenzene; **46** 5,5-dimethylcyclopent-2-enone;

48

49

Problems 11-30: *11* 1-phenylpropan-2-one; *12* propanamide; *13* 1,2-diphenylethane-1,2-dione (benzil); *14* 4-methoxy-4-methylpentan-2-one; *15* methyl 4-oxohexanoate; *16* 2-bromo-1,1-diethoxyethane; *17* 1-bromo-3-methylbutane; *18* (*E*)-ethyl but-2-enoate (ethyl crotonate); *19* (2-nitrophenyl)methanol; *20* 2-(3,5-dimethoxyphenyl)ethanenitrile; *21* 5-bromo-2-methylaniline; *22* (*E*)-3-(4-(dimethylamino)phenyl)prop-2-enal; *23* pyridin-2-amine; *24* 1-fluoro-4-nitrobenzene; *25* 2-chloro-5-nitrobenzoic acid; *26* benzyl 3-oxobutanoate; *27* pyridine-4-carbaldehyde; *28* 4-(4-methoxyphenyl)butanoic acid; *29* 4-methyl-2-nitrophenol; *30* 2-(phenoxymethyl)oxirane;

Problem 31   *Problem 32*

The curved arrows indicate the 2- and 3-bond correlations which are available from the HMBC spectrum.

Problem 33: 2-methyltetrahydrofuran

The names of compounds are exactly those derived from the Convert-Structure-to-Name programme in ChemDraw. They are all very simple, but if you are having problems interpreting any of them, you can type the name as shown here into a ChemDraw file, and the Convert-Name-to-Structure program will give you the structure.

For spectroscopic problems available on the Internet, see:
http://www.nd.edu/~smithgrp/structure/workbook.html

Index

CPSIA information can be obtained
at www.ICGtesting.com
Printed in the USA
LVHW101608100119
603457LV00013B/384/P

9 780077 11812